PROGRESS IN QUANTUM ELECTRONICS

Editors: J. H. Sanders and S. Stenholm

VOLUME 3 PART 3

THE QUANTUM THEORY OF THE LASER

THE QUANTUM THEORY OF THE LASER

BY

A. P. KAZANTSEV
L. D. Landau Institute for Theoretical Physics, Moscow

G. I. SURDUTOVICH
Semiconductor Physics Institute, Novosibirsk

PERGAMON PRESS

OXFORD · NEW YORK · TORONTO
SYDNEY · BRAUNSCHWEIG

PERGAMON PRESS LTD.
Headington Hill Hall, Oxford

PERGAMON PRESS INC.
Maxwell House, Fairview Park, Elmsford, New York 10523

PERGAMON OF CANADA LTD.
207 Queen's Quay West, Toronto 1

PERGAMON PRESS (AUST.) PTY. LTD.
19a Boundary Street, Rushcutters Bay, N.S.W. 2011, Australia

VIEWEG & SOHN GmbH
Burgplatz 1, Braunschweig

First edition 1974

Library of Congress Catalog Card No. 70-13569

Printed in Great Britain by Page Bros (Norwich) Ltd., Norwich

ISBN 0 08 017923 1

THE QUANTUM THEORY OF THE LASER

A. P. KAZANTSEV
L. D. Landau Institute for Theoretical Physics, Moscow

G. I. SURDUTOVICH
Semiconductor Physics Institute, Novosibirsk

CONTENTS

This review deals with the quantum theory of a single-mode laser. By using the small quasiclassical parameter $1/\sqrt{\bar{n}}$ (where $\bar{n}$ is the average number of photons), it is possible to obtain a closed set of quantum kinetic equations. This set of equations can be used to investigate lasers of any type at arbitrary radiation energies. The statistical properties of radiation of the main types of lasers are considered, and a comparison is made with the experimental data.

It is shown that the fluctuations of the photons at the critical point in a laser with nonlinear absorption are appreciably greater than in an ordinary laser; in a laser with a nonmonotonic dispersion contour, the frequency fluctuations may also become anomalously great.

The analogy with the theory of phase transitions of the first and second kind and with critical phenomena is investigated in detail.

1. INTRODUCTION

1.1. Fluctuations of Laser Radiation

The quantum theory of the laser is used to study the statistical properties of laser radiation due to spontaneous emission from atoms. Here we mean fluctuations of the radiation that are a result of the quantized structure of the electromagnetic field and of the active medium.

It should be emphasized that here we shall not consider technical noise (discharge noise, vibration of the reflectors of a resonator, etc.); these depend on external conditions and can sometimes be eliminated, or separated from the quantum fluctuations. Spontaneous emission from atoms determines a level of fluctuations that in principle cannot be reduced.

The quantum properties of the field and of the medium thus impose definite limitations on such coherence properties of radiation as the oscillation linewidth and the stability of the frequency, the level of the amplitude fluctuations, the correlation properties of the radiation over large space and time intervals, and the stability of the oscillation modes. Investigation of these limitations is of interest not only in principle, but also from a practical viewpoint.

Two approaches are possible in describing quantum noise. Within the framework of the Langevin noise theory, random forces may be introduced into the classical equations for the field and the medium. Such a method is effective, however, only at low radiation energies near the threshold (see subsection 1.5). The second approach is based on solving quantum equations for the field and the medium. In this case the statistical properties of the radiation are determined from first principles. This method makes it possible to consider fluctuations at any radiation energy.

The present review sets out a solution of the quantum problem for a single-mode laser (section 2). The quasi-classical parameter $1/\sqrt{\bar{n}}$, where $\bar{n}$ is the number of photons in a resonator, is the small parameter of the problem. The solution obtained by expansion with respect to $1/\sqrt{\bar{n}}$ may be considered as accurate for practical purposes, since $\bar{n}$ is very large.

Such an approach makes it possible to consider different kinds of lasers with arbitrary radiation energies by using a single method. On the basis of the quantum theory it is possible to give a simple physical picture of radiation fluctuations that is founded on the notion of the interaction of spontaneous electromagnetic wave trains. Section 3 deals with the basic statistical characteristics of a laser and gives their physical interpretation. Section 4 considers some applications of the theory. The analogy with phase transitions and critical phenomena is traced in detail.

The investigation of the fluctuations consists in finding the three main characteristics of radiation: (a) the fluctuations of the number of photons $\overline{\Delta n^2} = \overline{n^2} - \bar{n}^2$, (b) the radiation linewidths (phase fluctuations), and (c) the time correlation of photons $\overline{\Delta n(t)\,\Delta n(0)}$.

(a) *Fluctuations of photons*

The transition of a laser to self-sustained operation as an oscillator is a second-order phase transition. Figure 1 shows the dependence of the radiation energy on the normalized gain parameter η, equivalent to an equation of state of the system. Here $\eta = N/N_{th}$ where N is the number of excited atoms, and N_{th} is the threshold value of this number.

FIG. 1. Dependence of the intensity of radiation $|E|^2$ on the generation parameter. The solid line corresponds to the classical theory, the dotted line to the quantum theory.

Hence $\eta = 1$ corresponds to the threshold of oscillation when the field is described classically. The solid line shows the phase transition in an infinite volume. Such a dependence is given by the classical theory, which does not take fluctuations into consideration. The dotted curve corresponds to calculations according to the quantum theory and shows the phase transition in a system with a finite number of atoms and photons. When the finite nature of the number of particles is taken into account, this results, usually, in a blurring of the phase transition picture. The greatest difference between the quantum and the classical theories appears in the vicinity of the oscillation threshold.

Thus, at the threshold itself the average number of photons and the dispersion are of the order

$$\overline{\Delta n_{\text{th}}^2} \sim \bar{n}_{\text{th}}^2 \sim N_{\text{th}}. \tag{1.1}$$

Far from threshold the fluctuations approach the Poisson law $\overline{\Delta n^2} \sim \bar{n}$. The majority of theoretical and experimental investigations have studied the vicinity of the oscillation threshold, since the fluctuations here are greater and they are easier to observe.

A first-order phase transition is possible in a laser with nonlinear absorption, and the fluctuations in this are greater than in an ordinary laser. For example, at the critical point, i.e. the oscillation threshold, we have

$$\overline{\Delta n_{\text{th}}^2} \sim \bar{n}_{\text{th}}^2 \sim N_{\text{th}}^{\frac{4}{3}}. \tag{1.2}$$

In a gas laser this quantity is about two orders of magnitude larger than N_{th}. This question will be dealt with in greater detail in section 4.

(b) *Fluctuations of the phase*

The quantum phase fluctuations determine the width of the radiation line. To find this the coefficient of phase diffusion must be calculated.

Fluctuations of the phase $\Delta\varphi(t) = \varphi(t) - \varphi(0)$ are caused by fluctuations of the number of photons in a resonator, and so the diffusion coefficient can be found from the uncertainty relation $\Delta\varphi \cdot \Delta n \sim 1$. It should be noted that the characteristic time of photon fluctuations is $1/2\nu$, where ν is the resonator linewidth. In the region of classical oscillation $\Delta n \sim \sqrt{\bar{n}}$, so that diffusion of the phase is approximately expressed by the relation $\overline{\Delta\varphi^2} \sim \nu t/\bar{n}$. Hence for the linewidth $\Delta\nu$ we get the Townes formula

$$\Delta\nu \sim \nu/\bar{n}. \tag{1.3}$$

In a gas laser the typical value of the quantum width is about 0.1 Hz.

It must be noted that in a laser with a nonmonotonic dispersion contour, the fluctuations of phase may become anomalously large and at a certain critical point we have (subsection 4.2)

$$\Delta\nu \sim \nu/\bar{n}^{\frac{1}{3}}. \tag{1.4}$$

(c) *Time correlation of photons*

In the vicinity of the oscillation threshold the time correlation function of photons $G(t) = \overline{\Delta n(t)\,\Delta n(0)}$ determines the counting rate of paired coincidences of photons. The Fourier transform of $G(t)$ gives the frequency spectrum of the intensity fluctuations.

In a gas laser the correlation function is monotonically damped with time. The damping decrement reaches its minimum value at the threshold. It is still smaller (about $N_{\text{th}}^{\frac{1}{6}}$ times) in a laser with nonlinear absorption at the oscillation threshold i.e. the critical point.

In other cases $G(t)$ begins to oscillate with time, with a frequency which depends upon the ratio of the relaxation times of the medium and the field. In other words, a peak appears in the spectrum of the intensity fluctuations near the frequency of these oscillations.

1.2. The Model of a Laser

Let us consider the simplest model of a quantum generator. The active medium is a gas of excited atoms with a low density. We shall neglect the direct interaction between the atoms and take into account only their interaction with an electromagnetic field. We assume single-mode oscillation. In the resonance approximation it is sufficient to take into account only two states a and b of the atoms. Let us denote the frequency of the laser transition by ω_{ab} and the field frequency by ω_0.

We thus deal with a quantum oscillator having a frequency ω_0 that resonantly interacts with a system of N two-level atoms. The well-known spin Hamiltonian corresponding to such a model (the system of units in which $\hbar = 1$ is used) is

$$H = \omega_0 \overset{+}{a} a + \frac{1}{2}\omega_{ab}\sum_{i=1}^{N}\sigma_3^{(i)} + g\sum_{i=1}^{N}(a\overset{+}{\sigma}{}^{(i)} + \overset{+}{a}\sigma^{(i)}). \tag{1.5}$$

Here $\overset{+}{a}$ and a are the Bose operators of photon creation and annihilation which satisfy the commutation relation

$$a\overset{+}{a} - \overset{+}{a}a = 1. \tag{1.6}$$

The matrices $\sigma^{(i)}$ are the Pauli spin matrices for the ith spin

$$\sigma = \begin{pmatrix} 0 & 0 \\ 1 & 0 \end{pmatrix}, \quad \overset{+}{\sigma} = \begin{pmatrix} 0 & 1 \\ 0 & 0 \end{pmatrix}, \quad \sigma_3 = \begin{pmatrix} 1 & 0 \\ 0 & -1 \end{pmatrix} \tag{1.7}$$

where $\overset{+}{\sigma}$ is the creation operator of an atom in the upper state, and σ is the creation operator of an atom in the lower state.

The coupling constant g is expressed through the dipole moment of the transition d_{ab} and the quantization volume V by the relation

$$g = d_{ab}\sqrt{\left(\frac{\omega_0}{2V}\right)}. \tag{1.8}$$

It is assumed here for simplicity that the atoms are in identical conditions and for this reason all the coupling constants are the same. Such a situation is achieved for homogeneous broadening in the case of a travelling electromagnetic wave. In other cases the dependence of g on the coordinate of an atom may become significant.

The last simplification consists in that we consider the case of exact resonance: $\omega_{ab} = \omega_0$. Passing over to the interaction representation, we finally have

$$H = g\sum_{i}(a\overset{+}{\sigma}{}^{(i)} + \overset{+}{a}\sigma^{(i)}). \tag{1.9}$$

1.3. Relaxation and Excitation Processes

Let us consider the relaxation processes connected with the finite lifetimes of the atoms in the laser levels and of the photons in a resonator.

The usual way of describing relaxation consists in adding to eq. (1.9) the Hamiltonian of the interaction of the atoms and photons with their surroundings ("thermostats") that results in relaxation. To simplify this procedure, let us introduce the relaxation operators directly into the initial equation for R—the density matrix of our system "atoms + radiation".

The initial equation for the density matrix thus has the form

$$\frac{\partial R}{\partial t} = \hat{\nu}R + \sum_i \hat{\tau}_i^{-1}R + \frac{1}{\mathrm{i}}[H, R]. \tag{1.10}$$

The emergence of a photon from the resonator is a stochastic process, hence the relaxation of the field is described by the operator $\hat{\nu}$:

$$\hat{\nu}R = \nu(2aR\overset{+}{a} - \overset{+}{a}aR - R\overset{+}{a}a). \tag{1.11}$$

This operator can be obtained with the aid of the Hamiltonian (1.9) in second order perturbation theory. It corresponds to the process of linear absorption of photons (without taking account of saturation) by a certain number of atoms in the lower resonant state. Thus damping of the mean value of the field amplitude, which is proportional to the quantity $\bar{a} = \mathrm{Tr}\,(aR)$, in a free resonator follows the law

$$\bar{a}(t) = \bar{a}(0)\,\mathrm{e}^{-\nu t} \tag{1.12}$$

so that ν is the linewidth of the free resonator.

The relaxation operator $\hat{\nu}$ conserves the Hermitian character and the normalization of the density matrix:

$$\mathrm{Tr}_a(\hat{\nu}R) = 0 \tag{1.13}$$

where Tr_a is the trace with respect to the variables of the oscillator.

The relaxation and excitation of the ith atom are described by the operator $\hat{\tau}_i^{-1}$:

$$\hat{\tau}_i^{-1}R = (r_i^{(0)}\langle R\rangle_i - R)/\tau \tag{1.14}$$

where $r_i^{(0)}$ is the single-particle density matrix of the ith atom in the absence of radiation, its form depends on the level (upper or lower) to which the atom is excited initially. In the model which we have chosen all the atoms are excited to the upper laser level, so that $r_i^{(0)} = \overset{+}{\sigma}{}^{(i)}\sigma^{(i)}$. The symbol $\langle\ldots\rangle_i$ denotes the trace with respect to the states of the ith atom. It should be noted that the atomic relaxation operator (1.14) corresponds to the case when there is no correlation in the processes of excitation of different atoms, and, in addition, an excited atom is considered to be unpolarized; polarization appears only as a consequence of the field.

The relaxation operator (1.14) also conserves the normalization and Hermitian character of the density matrix R. For brevity we have written into it only one relaxation time τ. Under real conditions each state of an atom has its own lifetime τ_a and τ_b, and in addition there is the

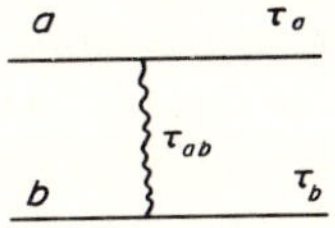

FIG. 2. Diagram of laser levels. τ_a = lifetime of upper level, τ_b = lifetime of lower level. τ_{ab} = inverse linewidth of radiation.

transverse relaxation time τ_{ab} (see Fig. 2). For this reason the right-hand side of eq. (1.14) should actually contain a matrix of τ_a^{-1}, τ_b^{-1} and τ_{ab}^{-1} instead of simply τ^{-1}. In the general case the operator $\hat{\tau}_i^{-1}$ does not conserve the normalization of the two active states a and b.

Various limiting cases are possible depending on the relation between the relaxation constants of the atoms and the resonator. There are only three cases, however, that differ in principle.

1.4 Three Types of Laser

Simplifying the picture somewhat, quantum generators can be divided into the following three characteristic groups (A), (B) and (C):

$$\nu\tau_a \ll 1, \qquad \nu\tau_b \ll 1, \qquad \nu\tau_{ab} \ll 1. \tag{A}$$

This case is characteristic of a gas laser. Thus, in a He–Ne laser the parameter $\nu\tau \sim 0.1$–0.01. Here we shall also consider the case of considerably different lifetimes of the levels a and b, namely, $\tau_b \ll \tau_a$.

In cases (B) and (C) the difference between τ_b and τ_a is not very appreciable, and therefore for simplicity we shall assume them to be equal.

$$\tau_a \approx \tau_b \approx \tau_{ab} \approx \tau, \qquad \nu\tau \gtrsim 1. \tag{B}$$

Case (B) is interesting in that when $\nu\tau > 2$ and the energy of radiation is sufficiently large, single-mode operation becomes unstable and passes over to an oscillating regime.

A molecular generator or maser has such parameters. Owing to the very large numerical value of the parameter $\nu\tau \sim 10^3$, however, at low energies the operation is stable (see the region of instability in Fig. 9).

In gas lasers the parameter $\nu\tau$ may apparently sometimes be slightly larger than unity. Such a situation is found in a mercury laser at 1.52 μm and a He–Ne laser at 3.39 μm at a low pressure, when broadening as a result of collisions becomes small.

$$\tau_a \approx \tau_b \approx \tau; \qquad \tau_{ab} \ll \tau; \qquad \nu\tau \gtrsim 1; \qquad \nu\tau_{ab} \ll 1. \tag{C}$$

This case is typical of solid-state, semiconductor and CO_2 lasers. Thus, in a ruby laser $\tau_{ab} \sim 10^{-12}$ s, $\tau \sim 10^{-3}$ s, and in a CO_2 laser $\tau_{ab} \sim 10^{-8}$ s, $\tau \sim 10^{-3}$ s. With such a ratio of relaxation times instability does not appear, but the level of fluctuations may be much higher than in case (A).

It should be noted that in a CO_2 laser with a nonlinear absorption cell, the effective parameter $\nu\tau_{ab}$ may become larger than unity with a low pressure in the cell, and we shall then be dealing with case (B).

In the above cases the fluctuations have different natures. In case (A) the spontaneous emission by separate atoms may be considered as statistically independent. In case (B) account must be taken of the interaction of the spontaneous electromagnetic wave trains emitted by different atoms. The interaction has a coherent nature. Finally, in case (C) account must also be taken of the interaction of the trains, but it has an incoherent nature.

1.5. Quantum and Quasiclassical Approach

Since a strong electromagnetic field is quasiclassical, in many investigations fluctuations have been studied within the framework of Langevin noise theory; random forces were introduced into the classical equations for the field and the medium in refs. [1]–[4].

In such an approach the problem consists of determining the correlation function of the random forces. This problem is simple only near the oscillation threshold, as long as $|\eta - 1| \ll 1$. Let us consider this question for the example of a gas laser (case A).

In this case the fluctuations of the complex amplitude of the field $E(t)$ are described by the van der Pol equation with the random force $f(t)$:

$$\frac{\partial E}{\partial t} = \chi(|E|^2)\,E + f(t) \tag{1.15}$$

$$\langle \overset{*}{f}(t)\,f(0)\rangle = D\delta(t)$$

where $\chi(|E|^2)$ is the total gain factor taking account of losses in the resonator.

Fluctuations of the field are connected with spontaneous emission from atoms and have a characteristic time scale τ. For this reason, owing to the condition that $\nu\tau \ll 1$, the random force correlation function can be replaced by a δ-function in time.

In the given case the field distribution function $F(tE)$ obeys the Fokker–Planck equation

$$\frac{\partial F}{\partial t} + \frac{\partial}{\partial E}(\chi E F) + \frac{\partial}{\partial \overset{*}{E}}(\chi \overset{*}{E} F) = D\frac{\partial^2 F}{\partial E\,\partial \overset{*}{E}}. \tag{1.16}$$

To determine the diffusion coefficient D let us consider a laser below oscillation threshold.

Below the threshold we can introduce the effective temperature of radiation T with the aid of the relation

$$N/N_{\text{th}} = \mathrm{e}^{-\omega_0/k_B T} \approx 1 - \omega_0/k_B T.$$

The temperature is expressed through the gain parameter $k_B T = \omega_0/(1-\eta)$. Hence, in particular, we can find the average number of photons below threshold $\eta < 1$:

$$\bar{n} = (1 - \eta)^{-1}. \tag{1.17}$$

The stationary function in this case coincides with the distribution for black-body radiation

$$F = \text{const.} \exp\left(-\frac{|E|^2 V}{4\pi k_B T}\right). \tag{1.18}$$

Below threshold the field is very weak. For this reason nonlinear effects are not important and $\chi \approx \nu(\eta - 1) < 0$. To obtain the distribution (1.18) for F, we must assume that

$$D = 4\pi\nu\omega_0/V. \tag{1.19}$$

It is quite clear that the random forces have no special features for small values of the parameter $1 - \eta$. For this reason the values of D found below the threshold may also be used above the threshold as long as $|\eta - 1| \ll 1$. It should be noted that eq. (1.16) is widely used in the theory of electronic generators, small-size superconductors near the critical temperature and in other fields of physics. This equation describes the simplest model of a second-order phase transition in a small volume. The description of the threshold region thus has universal validity; the field fluctuations can be determined here from phenomenological considerations.

When $\eta - 1 \gtrsim 1$ the intensity of the random forces cannot be determined from thermodynamic considerations. In this case the dependence of D on η must be taken into consideration. For progress in this field, it is necessary to solve the quantum problem of the spontaneous emission of atoms in a strong field. Such an approach, based on the solution of quantum equations of motion for a field and a medium, was developed in a number of papers [5–11].

The problem of spontaneous emission from one free atom in a strong field was considered with certain simplifications in refs. [12]–[14]. It must be stressed that in the present case we are interested only in the part of the spontaneous emission of an atom that contributes to the selected field mode. It can be understood that such a contribution in a volume much larger than the wavelength is quite small. The final effect appears only because of the large number of atoms.

Scully and Lamb developed the quantum theory of a gas laser. Of appreciable significance was the condition $\nu\tau \ll 1$ which made it possible to neglect correlation in the spontaneous emission of different atoms. Having calculated the interaction of one excited atom with a quantized electromagnetic field (it is assumed that the field density matrix has not time to change during the time τ) and multiplied the result by the number of atoms, the authors obtained a quantum kinetic equation for P_n—the probability of detecting n photons in a resonator:

$$\frac{\partial P_n}{\partial t} = Q_n - Q_{n+1}$$

$$Q_n = a(n)P_{n-1} + b(n)P_n \tag{1.20}$$

This equation is a discrete analogue of the Fokker–Planck equation (1.16). Although the discrete form of writing the equation is quite symbolic here (owing to the large value of n), it is very illustrative from a physical viewpoint. The absence of correlation in the emission processes results in the fact that the probability flux between the n and $n - 1$ levels of a quantum oscillator Q_n depends only on P_n and P_{n-1}. In other words, there are no correlated pairs, triplets, etc., of photons in the spontaneous emission.

When the condition $\nu\tau \ll 1$ does not hold, correlation appears in the spontaneous emission from different atoms. If we attempt to include corrections of higher order in the parameter $\nu\tau$ in eq. (1.20), then terms with P_{n+1}, P_{n-2}; P_{n+2}, P_{n-3} etc., that will be quadratic, cubic, etc. with respect to the density of the emitting atoms, will appear in the expression for the flux Q_n. These additional terms correspond to correlated pairs and triplets of photons in the spontaneous emission.

A quantum kinetic equation for P_n with an arbitrary ratio of the relaxation times of atoms and photons was obtained by Schwinger [15]. It is not local in time, but shows memory effects. This equation, however, takes into account only effects linear in the field, and for that reason it cannot be directly applied to a laser.

In the following section we shall consider a solution of the quantum problem obtained by an expansion with respect to the natural small parameter $1/\sqrt{\bar{n}}$. This will allow us to consider the statistical properties of laser radiation at arbitrary energies and with any relaxation constants.

2. THE EXACT SOLUTION OF THE QUANTUM PROBLEM

2.1. Coherent States

It is not convenient to use the representation of occupation numbers in the quasiclassical region. It is much more natural to work using the coherent state representation; here it is easy to separate the large mean value of the field from the small fluctuating part. The properties of coherent states have been thoroughly investigated (see, for example, refs. [16]–[19]). Let us recall the basic relations and introduce the definitions that will be used in the subsequent theory.

The coherent vector $|z\rangle$ is the eigenstate of the annihilation operator:

$$a|z\rangle = z|z\rangle, \quad z = x + \mathrm{i}y \tag{2.1}$$

where z is a complex parameter.

The relation of $|z\rangle$ to the state vectors with a fixed number of photons is given by the expression

$$|z\rangle = \mathrm{e}^{-|z|^2/2} \sum_{n=0}^{\infty} \frac{z^n}{\sqrt{n!}} |n\rangle. \tag{2.2}$$

Since the operator a is non-Hermitian, the different vectors $|z\rangle$ and $|z'\rangle$ are not orthogonal

$$|\langle z'|z\rangle|^2 = \mathrm{e}^{-|z-z'|^2}. \tag{2.3}$$

Let us introduce the density matrix ρ of the quantum oscillator. It is very convenient to use the Glauber–Sudarshan diagonal representation for ρ, namely,

$$\rho = \int \mathrm{d}^2 z \rho(z) \, |z\rangle \, \langle z| \tag{2.4}$$

$$\mathrm{Tr}_a(\rho) = \int \mathrm{d}^2 z \rho(z) = 1. \tag{2.5}$$

Integration is performed over the entire complex plane z.

The function $\rho(z)$ is called a weight function; in the quasiclassical region it is close to the classical field distribution function $F(tE)$, but generally speaking differs from it.

In the following we shall also use polar coordinates in the z-plane:

$$z = (\sqrt{\xi})\, e^{i\varphi}; \qquad \overset{*}{z} = (\sqrt{\xi})\, e^{-i\varphi} \tag{2.6}$$

and the Fourier series expansion of $\rho(z)$ with respect to the angle:

$$\rho(z) = \frac{1}{\pi} \sum_{m=-\infty}^{+\infty} e^{-im\varphi} \rho_m(\xi); \qquad \overset{*}{\rho}_m = \rho_{-m}. \tag{2.7}$$

The latter relation ensues from the fact that the weight function is real.

In the expansion (2.7) the quantity $\rho_0(\xi)$ plays the part of the distribution function of the number of photons. Indeed, let us determine the average number of photons $\bar{n}$ and the dispersion $\overline{\Delta n^2}$ in the usual way:

$$\begin{aligned} \bar{n} &= \mathrm{Tr}_a(\overset{+}{a} a \rho) \\ \overline{\Delta n^2} &= \mathrm{Tr}_a((\overset{+}{a} a)^2 \rho) - \bar{n}^2. \end{aligned} \tag{2.8}$$

These quantities are expressed through the function $\rho_0(\xi)$ as follows:

$$\bar{n} = \int_0^\infty d\xi\, \xi \rho_0(\xi); \qquad \int_0^\infty d\xi \rho_0(\xi) = 1 \tag{2.9}$$

$$\overline{\Delta n^2} = \int_0^\infty d\xi\, \xi^2 \rho_0(\xi) - \bar{n}^2 + \bar{n}. \tag{2.10}$$

The presence of the last term on the right-hand side of eq. (2.10) makes $\rho_0(\xi)$ differ from the usual distribution function. Hence $\rho_0(\xi)$ coincides with the classical distribution function of photons only when $\overline{\Delta n^2} \gg \bar{n}$.

The weight function is not always a positive quantity. Thus, for the n-quantum state with a density matrix $\rho = |n\rangle\langle n|$ the weight function has the form $\rho(z) \sim e^{\xi} \delta^{(n)}(\xi)$ where $\delta^{(n)}$ is the nth derivative of the δ-function.

In the following we shall not consider pure states nor other cases where the weight function is too singular and the use of the $|z\rangle$-representation instead of the $|n\rangle$-representation gives no advantages.

We shall be interested only in the quasiclassical situation, which we shall define by two conditions:

$$\bar{n} \gg 1 \tag{2.11}$$

$$\overline{\Delta n^2} - \bar{n} \gg 1. \tag{2.12}$$

The meaning of the second condition is as follows. In the quasiclassical region quantum corrections can be obtained in the form of an expansion with respect to the small parameter

$$\frac{1}{\rho}\frac{\partial \rho}{\partial \xi} \ll 1. \tag{2.13}$$

This is the small parameter of the problem. In the classical region (above the oscillation threshold), the function $\rho_0(\xi)$ is Gaussian

$$\rho_0(\xi) = (2\pi d\bar{n})^{-\frac{1}{2}} \exp\left[-(\xi - \bar{n})^2/2d\bar{n}\right] \tag{2.14}$$

with the dispersion parameter d characterizing the deviation of fluctuations from Poisson's law and related to $\overline{\Delta n^2}$ by the expression

$$\overline{\Delta n^2} = (1 + d)\bar{n}. \tag{2.15}$$

Consequently the small parameter has the following form:

$$\frac{1}{\rho^0}\frac{\partial \rho_0}{\partial \xi} \sim (d\bar{n})^{-\frac{1}{2}} \ll 1. \tag{2.13'}$$

We thus consider the function $\rho_0(\xi)$ to be sufficiently broad near $\bar{n}$. Hence condition (2.12) follows.

It should be stressed that the condition (2.12) does not imply any untenable assumption. In practice it is almost always obeyed. Otherwise we would have to deal with a radiating system in which fluctuations are absent. Such a system is called a predetermined current; in this case the electromagnetic field is in a strictly coherent state and $\overline{\Delta n^2} = \bar{n}$. For this reason condition (2.12) implies that we are dealing with a sufficiently broad packet of coherent states. In a real situation, owing to the great number of atoms, fluctuations of the medium always ensure that the condition (2.12) is satisfied.

As regards the harmonics ρ_m of the weight function, the averages of the operators of the type a^m, $\overset{+}{a}{}^l a^{m+l}$, etc., are expressed through them. Thus, the average value of the positive frequency part of the field amplitude is proportional to the quantity

$$\bar{a} \equiv \mathrm{Tr}_a(a\rho) = \int_0^\infty d\xi(\sqrt{\xi})\rho_1(\xi). \tag{2.16}$$

The field phase φ plays the part of the order parameter. In a finite volume order cannot exist for an unlimited time (the Onsager theorem). In the present case this means that in conditions of steady oscillation all the harmonics except for $\rho_0(\xi)$ damp out with time. Therefore under steady conditions we have

$$\rho_1(t\xi) = \rho_1(0, \xi)\, \mathrm{e}^{-\Delta\nu t}. \tag{2.17}$$

Scully and Lamb showed [20] that $\Delta\nu$ is the linewidth of the laser radiation. In other words, instead of the correlation function determining the radiation linewidth

$$\overline{\overset{+}{a}(t)\, a(0)} \sim \mathrm{e}^{-\Delta\nu t} \tag{2.18}$$

it is sufficient to determine the damping of the average field $\bar{a}(t)$.

2.2. Equations of Motion in the Coherent State Representation

Let us now go over to the derivation of the equations of motion for the density matrices of a quantum oscillator and atoms in the representation of coherent states.

For brevity we shall omit the relaxation operators. It is easy to restore them in the final equations. However, in order not to lose the relation to the stationary situation, which appreciably depends on the ratio of the parameters ν and τ, it must be noted that in a condition of steady oscillation the following balance equation holds:

$$2\nu\bar{n} = (N - N_{th})/\tau. \tag{2.19}$$

Hence it can be seen that the inequality $\bar{n} \gg N$ when $\eta - 1 \sim 1$ corresponds to the case $\nu\tau \ll 1$, i.e. the energy of the system is mainly concentrated in the electromagnetic field. Consequently in this case the problem of the interaction of the field with the atoms can be considered in the approximation of the predetermined field—initially the given field energy is not changed noticeably by the interaction.

When $\nu\tau \lesssim 1$ we have $\bar{n} \gtrsim N$ and the problem must be considered in a self-consistent approximation for the field, since the energy of the field changes appreciably during the interaction with the atoms. Thus the inequality $N \lesssim \bar{n}$ is the criterion for self-consistency of the field.

To go over to the $|z\rangle$-representation, let us write the density matrix of the entire system $R(t)$ in diagonal form (2.4):

$$R(t) = \int d^2 z \, R(tz) \, |z\rangle \, \langle z| \tag{2.20}$$

$$\mathrm{Tr}\,(R(t)) = \int d^2 z \langle R(tz) \rangle_{1, 2, \ldots, N} = 1. \tag{2.21}$$

Here, and below, the symbol $\langle \ldots \rangle_{1, 2, \ldots, N}$ denotes the trace with respect to the variables of the 1st, 2nd, ..., Nth atoms.

Let us now introduce eq. (2.20) into the equation for the density matrix:

$$\mathrm{i} \frac{\partial R}{\partial t} = [H, R] \tag{2.22}$$

$$H = g \sum_i (\overset{+}{a} \sigma^{(i)} + a \overset{+}{\sigma}{}^{(i)}). \tag{2.23}$$

Let us multiply the left-hand and right-hand parts of eq. (2.22) by an arbitrary vector $|z'\rangle$ and use the relation of orthogonality (2.3). After integration by parts we get equations for $R(tz)$

$$\mathrm{i} \frac{\partial R}{\partial t} = \mathscr{H} R - R \overset{+}{\mathscr{H}}, \qquad \mathscr{H} = \mathscr{H}_0 + \mathscr{H}_1 \tag{2.24}$$

$$\mathscr{H}_0 = \sum_i h_i, \qquad h_i = g(\overset{*}{z} \sigma^{(i)} + z \overset{+}{\sigma}{}^{(i)}) \tag{2.25}$$

$$\mathscr{H}_1 = -g \sum_i \sigma^{(i)} \nabla; \qquad \nabla = \frac{1}{2} \left(\frac{\partial}{\partial x} - \mathrm{i} \frac{\partial}{\partial y} \right). \tag{2.26}$$

In the new representation the Hamiltonian $\mathscr{H}$ is of a non-Hermitian nature, since the transformation to the $|z\rangle$-representation is not unitary.

The Hermitian part of the Hamiltonian $\mathscr{H}_0$ obviously describes the interaction of the atoms with the classical field z. Here only a precession of the atomic polarization vector occurs, and the energy of the field does not change. The non-Hermitian part of the Hamiltonian describes the processes of emission and absorption of photons in the system.

At a first glance, with respect to the orders of magnitude we have $\mathscr{H}_1/\mathscr{H}_0 \sim 1/\xi \sim 1/\bar{n}$. Within the quasiclassical limit $\bar{n} \to \infty$ it is therefore natural to attempt to consider $\mathscr{H}_1$ as a small perturbation. Actually this is possible only if $\xi \gg N$, when the approximation of the predetermined field holds. At $\xi \gtrsim N$ the perturbation theory becomes incorrect, since $\mathscr{H}_1$ also describes the effects of self-consistency of the field.

Indeed, in the quasiclassical region the density matrix of the system $R(tz)$ in the leading order can be expressed through the density matrices of the separate atoms $r_i(tz)$ and the weight function of the field $\rho(tz)$ in the factorized form:

$$R(tz) \approx \rho(tz)\, r_1(tz)\, r_2(tz) \ldots r_N(tz). \tag{2.27}$$

Comparing the results of $\mathscr{H}_0$ and $\mathscr{H}_1$ operating on the density matrix (2.27), we see that by operating on the atomic variables we have the ratio $\mathscr{H}_1/\mathscr{H}_0 \sim N/\xi$, and by operating on the oscillator variables $\mathscr{H}_1/\mathscr{H}_0 \sim \rho^{-1}/\partial\rho/\partial\xi)$.

The Hamiltonian $\mathscr{H}_1$ thus describes effects of two kinds: field self-consistency effects of the order $N/\bar{n}$ and effects of quantum diffusion in plane z of the order $1/\sqrt{\bar{n}}$.

In the following the field self-consistency effects will be taken into consideration accurately, and the quantum diffusion effects according to perturbation theory.

2.3. Derivation of a Closed Set of Equations

The representation of the density matrix in the form (2.27) is inaccurate, since it takes no account of correlation effects. To introduce these consecutively, let us form a Bogolyubov chain of equations for the first basic distribution functions.

We shall define these functions as follows:

$$\rho(tz) = \langle R(tz)\rangle_{1,2,\ldots,N}$$

$$\rho(tz)\, r_1(tz) = \langle R(tz)\rangle_{2,3,\ldots,N} \tag{2.28}$$
$$\rho(tz)\, r_{12}(tz) = \langle R(tz)\rangle_{3,4,\ldots,N}$$

Here r_1 and r_{12} are one- and two-particle density matrices of atoms. When written in such a form they have the correct normalization

$$\langle r_1\rangle_1 = 1; \quad \langle r_{12}\rangle_2 = r_1. \tag{2.29}$$

In the classical limit r_1 does not depend on ρ and the correlation effects are absent: $r_{12} = r_1 r_2$. If quantum effects are taken into account, then r_1 depends on $\nabla\rho/\rho$ and a small correlation correction appears in r_{12}.

To obtain equations for ρ, r_1 and r_{12}, a trace should be taken with respect to N, $N - 1$ and $N - 2$ variables in eq. (2.24). Since in our model all the atoms are identical, summation over the atoms consists in multiplication by N.

The equation for $\rho(tz)$ is an equation of continuity written in complex form:

$$\frac{\mathrm{d}\rho}{\mathrm{d}t} + \frac{1}{\mathrm{i}}(\nabla v - \mathrm{c.c.})\,\rho = 0$$
$$\frac{\mathrm{d}}{\mathrm{d}t} = \frac{\partial}{\partial t} + \frac{1}{\mathrm{i}}(v\nabla - \mathrm{c.c.}); \qquad v = gN\langle\sigma r\rangle \tag{2.30}$$

The complex radiation current v is similar to the rate of flow of a compressible liquid with density $\rho(tz)$ in the z-plane. It is quite clear that d/dt plays the part of a substantial derivative.

The equation for a single-particle density matrix has the form

$$\frac{1}{\mathrm{i}}\frac{\mathrm{d}r_1}{\mathrm{d}t} + [h_1, r_1] = g(\tilde{\sigma}^{(1)} r_1 + N\langle\sigma^{(2)}\delta r_{12}\rangle_2)\nabla \ln\rho - \mathrm{h.c.} \tag{2.31}$$

where $\tilde{\sigma} = \sigma - \langle\sigma r\rangle$.

Here a two-particle correlation matrix δr_{12} with a zero trace has been introduced in accordance with the relation:

$$r_{12} = r_1 r_2 + \delta r_{12}; \qquad \langle\delta r_{12}\rangle_1 = \langle\delta r_{12}\rangle_2 = 0.$$

Equation (2.31) also has a hydrodynamic form; it describes the motion of a spinor liquid unevenly distributed over the z-plane in the external field. The ratio of the right-hand side of the equation to $[h_1, r_1]$ gives the small parameter of the problem (2.13).

It should be noted that the trace of the right-hand side of eq. (2.31) is equal to zero, since $\langle\tilde{\sigma}\rangle = 0$. From a physical viewpoint this is connected with the fact that the fluctuations of the field at the operating frequency of the transition do not significantly change the probability of populating the upper and lower laser levels. For this reason the normalization of the single-particle density matrix $\langle r_1\rangle_1$ is conserved.

The set of eqs. (2.30) and (2.31) is not closed because of the correlation function δr_{12}. It is important that δr_{12} enters the last equation in the form of a small quantum correction. For this reason, when calculating δr_{12}, we need not take account of quantum corrections of the order $\rho^{-1}(\partial\rho/\partial\xi)$. Discarding such small terms, we get for δr_{12} an equation containing no new correlation functions.

$$\frac{1}{i}\frac{d\delta r_{12}}{dt} + [h_1 + h_2, \delta r_{12}] = gN\langle\sigma^{(3)}(\delta r_{13}\nabla r_2 + \delta r_{23}\nabla r_1)\rangle_3 + g(\tilde{\sigma}^{(1)}r_1\nabla r_2 + \tilde{\sigma}^{(2)}r_2\nabla r_1) - \text{h.c.} \quad (2.32)$$

We now have a closed set of eqs. (2.30)–(2.32) for the weight function ρ, the single-particle density matrix r_1 and the correlation density matrix δr_{12}. It should be emphasized that the set could be closed only thanks to the smallness of the quasiclassical parameter (2.13).

From the last equation it is easy to estimate the order of the correlation function

$$\delta r_{12} \sim r_1 r_2/\xi. \quad (2.33)$$

In a strong field the correlation function diminishes in inverse proportion to the number of photons. Such a dependence can be understood from general considerations. Indeed, a statistical dependence appears in the spontaneous emission of atoms as a result of the following process. A certain atom emits a photon, another atom absorbs it, and then emits it again. Then, if the first atom interacts with this photon, a certain correlation appears between the two atoms. The probability of the first atom interacting with this photon is $1/\bar{n}$. This agrees with the result of (2.33)†.

† In the following the correlation in the spontaneous emission of atoms will be interpreted as the interaction of spontaneous electromagnetic trains. In this way it is easy to consider the correlation effects with an arbitrary ratio of the relaxation constants.

Thus the correlation in the spontaneous emission of two atoms in a strong field is small. Such a correlation, however, extends over the entire volume occupied by the field and the medium. For this reason the correlation function δr_{12} enters into eq. (2.31) with a factor N and the significance of the correlation effects is the same as the criterion of self-consistency of the field $N \gtrsim \bar{n}$.

Neglecting quantum fluctuations, we can use eqs. (2.30) and (2.31) to obtain a classical Maxwell equation for the average field $\bar{z}(t)$ and an equation for the medium in this field. This simplest self-consistent system of equations was investigated in refs. [21]–[23].

The system of kinetic eqs. (2.30)–(2.32) was obtained in ref. [11]. In deriving these equations we did not have in mind any definite model of a laser. Consequently, the approach considered may be used for any type of laser. For this purpose it is sufficient to introduce the corresponding relaxation operators on the left-hand side of the equations. It is also easy to apply this approach in the case of Doppler broadening, a spatial structure of the medium being studied, etc.

It must be stressed that the set of kinetic equations obtained, when applied to a laser, holds not only on the quasiclassical region above the oscillation threshold, but also at and below the threshold. Below the threshold $h < 1/\tau$ and for this reason the small parameter of the quantum problem, instead of eq. (2.13), now has the form $g\tau\nabla \ln \rho \sim g\tau/\sqrt{\bar{n}}$ where $\bar{n}$ is determined from formula (1.17).

When $\bar{n} \sim 1$ the small parameter coincides with $g\tau$ and has approximately the same order of magnitude as eq. (2.13) (see formula (3.6) expressing the average number of photons through $g\tau$). Only at a very small intensity, $\bar{n} \sim (g\tau)^2$, is the assumption considered above violated. Such low intensities, however, are of no interest in the theory of the laser.

3. THE STATISTICAL PROPERTIES OF THE RADIATION

Let us now consider the statistical properties of the radiation with the aid of the set of kinetic equations including the relaxation operators obtained above. It will be convenient to study separately the limiting cases (A), (B) and (C) indicated in subsection 1.4.

3.1. The Gas Laser

In a gas laser the coherence parameter $\nu\tau$ is small and atoms emit independently of each other. The correlation matrix δr_{12} on the right-hand side of eq. (2.31) may therefore be omitted. Since the field changes slowly during a time of the order $1/\nu$, eq. (2.31) may be used in a quasistationary approximation, discarding the derivative with respect to time. Omitting the subscript 1 from eq. (2.31) we get

$$(r - r^{(0)})/\tau = \frac{1}{i}[h, r] + ig(\tilde{\sigma} r \nabla \ln \rho - \text{h.c.}). \tag{3.1}$$

This equation describes correlation between the states of an atom and the quantum oscillator. Let us explain the physical meaning of the operator $\tilde{\sigma}$ in this equation.

Let us recall that σ is the operator of the radiation current. Consequently, $\tilde{\sigma}$ is the operator of the fluctuation of the radiation current. The factor $\nabla\rho$ in eq. (3.1) determines the diffusion coefficient of the field in the z-plane in the equation for the weight function $\rho(tz)$. The diffusion coefficient is expressed through a certain average value of the square of the operator of the current fluctuation $\tilde{\sigma}$. Here matters are exactly the same as in the case of a Brownian particle in an external field. If $x(t)$ is a random coordinate of a particle, then the diffusion coefficient is determined by the mean value

$$\langle (x(t) - \langle x(t) \rangle)^2 \rangle / 2t. \tag{3.2}$$

The field diffusion factor is expressed similarly through $\tilde{\sigma}$. The square of the average current $\langle \sigma r \rangle$ determines the probability of stimulated radiation W_{st}, and the square of the current fluctuation determines the probability of spontaneous radiation W_{sp} [83]. Here $W_{st} + W_{sp} = 1$.

The diffusion coefficient for photons is proportional to W_{sp}. The probabilities W_{sp} and W_{st} are shown in Fig. 3 as functions of the field intensity. The value of W_{st} is low in a weak and a strong field (owing to saturation). This is why W_{sp} has a minimum at a saturating field E_0. In a more complicated model of relaxation (with several different relaxation times), the total probability of spontaneous and stimulated radiation is not conserved, so that the drop of W_{sp} may decrease. The difference between $\tilde{\sigma}$ and σ may thus become appreciable only at average energies of radiation $\eta - 1 \sim 1$.

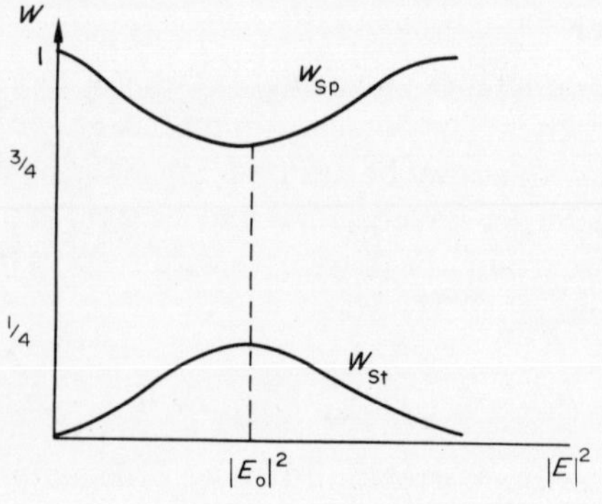

FIG. 3. Dependence of the square of the average radiation current (W_{st}) and the square of the current fluctuations (W_{sp}) on the field intensity. $|E_0|^2$ = characteristic intensity determined by the saturation parameter.

If in addition to the laser levels a and b other states must also be taken into consideration in an atom, for example the ground state, the matrix r is of a higher dimension and the trace with the respect to the two laser levels $\mathrm{Tr}_{a,b}(r) = r_{aa} + r_{bb}$ is less than unity. In this case $\tilde{\sigma}$ has the form

$$\tilde{\sigma} = \sigma - \frac{\mathrm{Tr}_{a,b}(\sigma r)}{\mathrm{Tr}_{a,b}(r)} \tag{3.3}$$

Here all the matrices relating to levels a and b should be considered multiplied by a projection operator equal to 1 when operating on the states a and b and to 0 in all the remaining cases.

Owing to definition (3.3) the trace of the right-hand side of eq. (3.1) with respect to the laser states a and b is equal to zero. This is connected with the fact that the interaction of an atom with the resonance field does not significantly change the population of levels a and b.

Consequently, eq. (3.1) is a linear one with respect to the normalization of the active states. This is why the result depends only on the population of the laser levels. Below, as previously, we shall consider the atoms to have only two levels.

It should be noted that the Scully and Lamb equation for the joint density matrix of an atom and the field in the coherent state representation can be reduced to the form of (3.1), except for the fact that it contains σ instead of $\tilde{\sigma}$. This difference is apparently connected with the circumstance that in the Scully and Lamb theory the equation for the joint matrix has been obtained in the approximation of an infinitely long lifetime of a photon in the resonator. The change in the state of the field during the lifetime of an atom is neglected. This approximation is undoubtedly sufficient for finding the classical radiation current. However, to find the quantum fluctuations, generally speaking, it is necessary to take into account the finite nature of the lifetime of a photon in the resonator, since the fluctuations of photons caused by their emission from the resonator is of approximately the same magnitude as those due to the spontaneous emission from the atoms.

Equation (3.1) is easy to solve by the use of perturbation theory. Inserting the result of such an approximate solution into the continuity equation (2.30), we get the Fokker–Planck equation for $\rho(t\xi\varphi)$

$$\frac{\partial \rho}{\partial t} = 2\nu \left\{ \frac{\partial}{\partial \xi}\left(A\rho + B\frac{\partial \rho}{\partial \xi} \right) + \frac{C}{\xi^2}\frac{\partial^2 \rho}{\partial \varphi^2} \right\} \tag{3.4}$$

$$A = \xi[1 - \eta(1 + 4g^2\tau^2\xi)^{-1}], \qquad B = \xi/\eta^2, \qquad C = \xi(\eta + 1)/8.$$

Here $(-\nu A/\xi)$ is the gain coefficient, and B and C are the radial and azimuthal diffusion coefficients.

The photon distribution

The stationary weight function $\rho_0(\xi)$ near the oscillation threshold has the form

$$\rho_0(\xi) = \text{const.} \exp\left[(\eta - 1)\xi - \beta\xi^2/2\right] \tag{3.5}$$

where $\beta = 4g^2\tau^2$ is the saturation parameter. This parameter is very small. In a gas laser $\beta \sim 10^{-7}$–10^{-8}.

Below the oscillation threshold, when $1 - \eta \ll \sqrt{\beta}$, eq. (3.5) changes into the Planck distribution, with the average number of photons (1.17):

$$\rho_0(\xi) = \frac{1}{\bar{n}} e^{-\xi/\bar{n}}; \qquad \bar{n} = (1 - \eta)^{-1}$$

$$\overline{\Delta n^2} = \bar{n}(\bar{n} + 1) \approx \bar{n}^2. \tag{3.5'}$$

Above the threshold when $\eta - 1 \gg \sqrt{\beta}$ the relative fluctuations are small and we have the case of classical oscillation. The transition from the Planck distribution to the distribution corresponding to classical oscillation takes place within the narrow interval $|\eta - 1| \sim \sqrt{\beta}$. Figure 4 shows the dispersion of the photons in the transition region. The distribution function of the photons in the vicinity of the oscillation

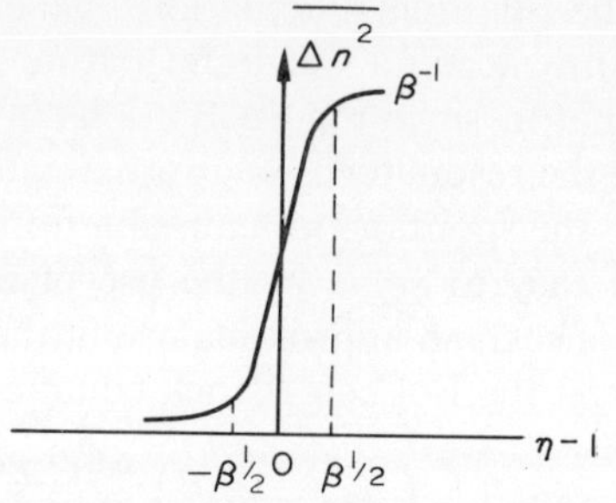

FIG. 4. Dispersion of photons in the vicinity of the oscillation threshold. The transition from thermal radiation to the region of classical oscillation takes place in the interval $|\eta - 1| \sim \sqrt{\beta}$.

threshold was investigated by photocounting techniques in refs. [24]–[27]. Not only was the dispersion of the photons measured, but also the third and fourth moments of the distribution function (3.5). The results of the measurements agree well with the calculations.

Above the oscillation threshold when $\eta - 1 \gg \sqrt{\beta}$ we have the quasiclassical distribution function (2.14). The average number of photons and the dispersion have the form

$$\bar{n} = (\eta - 1)/\beta; \qquad \overline{\Delta n^2} = (\eta - (\eta - 1)/\eta)/\beta \tag{3.6}$$

The second term in the expression for the dispersion appears owing to the difference between $\tilde{\sigma}$ and σ. As can be seen from the above, the largest contribution from this correction takes place at $\eta = 2$ and is 25%.

When $\tau_b \ll \tau_a$ the expression for the dispersion changes somewhat:

$$\overline{\Delta n^2} = \left(\eta - \frac{1}{2\eta}(\eta - 1)\right)\Big/\beta. \tag{3.7}$$

The behaviour of the dispersion above the oscillation threshold is shown qualitatively in Fig. 5. The straight dotted line corresponds to the result $\overline{\Delta n^2} = \eta/\beta$, curve 1 corresponds to formula (3.6), and curve 2 to formula (3.7).

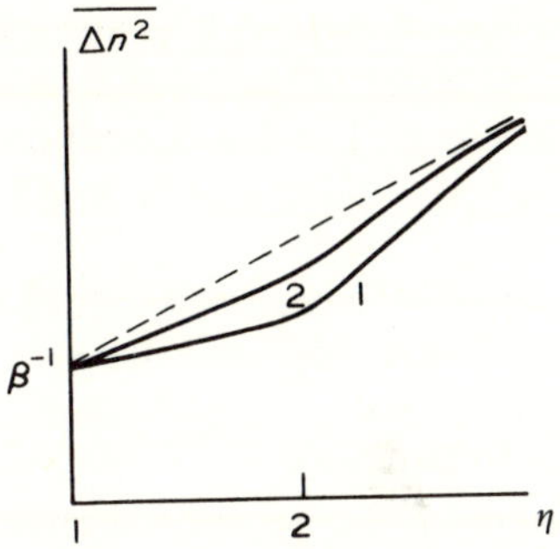

FIG. 5. Dispersion of photons above the oscillation threshold ($\eta - 1 \gg \sqrt{\beta}$). The straight dotted line corresponds to the equation $\overline{\Delta n^2} = \eta/\beta$. Curves 1 and 2 correspond to equations (3.6) and (3.7).

Thus, as a result of substituting σ for $\tilde{\sigma}$ the fluctuations of the photons in the region of intermediate energies decrease somewhat.

The formulas for the more general case taking account of the finite detuning and transitions of the atoms to the lower active state are given in Appendix I.

Photocounting techniques

An effective method for measuring the moments of the distribution functions of photons and their correlation properties is the use of photocounting techniques.

The statistics of the photocounts differs from that of the photons inside the resonator. The distributions are simply related to each other, however, so that knowing the moments of the distribution function of the photocounts, we can deduce the moments of distribution of photons in a laser. For example, the first two moments of the distribution of the photoelectrons in a photodetector $\overline{n^1}$ and $\overline{(\Delta n^1)^2}$ are related to the average number and dispersion of photons in a laser by the expression

$$\overline{n^1} = \varepsilon\bar{n}; \qquad \overline{(\Delta n^1)^2} = \varepsilon(1-\varepsilon)\bar{n} + \varepsilon^2\overline{\Delta n^2}. \tag{3.8}$$

The parameter ε is determined by a number of factors, namely, the transmission of the resonator mirrors, the attenuation of the intensity with a light filter, the masking of the photon beam and the quantum efficiency of the photodetector. The theory of photocounting techniques has been worked out in detail and described in refs. [24], [25], [28–30].

The fluctuation of the radiation under transient conditions

A number of authors have investigated theoretically and experimentally the properties of laser radiation under transient conditions, during the onset of steady-state oscillation.

A typical formulation of the problem consists in the following. Initially the laser is below threshold and has a gain parameter $\eta_1 < 1$. The statistics of the field are determined by the Planck distribution with an average number of photons $\bar{n} = (1-\eta_1)^{-1}$. Then, by rapidly changing the quality factor of the resonator, the laser is transferred to a state above threshold with the parameter $\eta_2 > 1$. After this there is a transient process in which steady oscillation with $\eta = \eta_2$ is finally achieved.

The process of transition to a steady state can be divided into two stages. In the first initial stage amplification of the thermal noise which existed before the moment of triggering takes place, and also amplification of the spontaneous emission, which occurs after triggering. The field distribution function $\rho(\xi t)$ coincides as previously with the Planck distribution (3.5′) and $\bar{n}$ depends on time according to the law

$$\bar{n}(t) = \bar{n}_0(t)\,e^{t/t_0}; \qquad \bar{n}_0(t) = (1-\eta_1)^{-1} + (\eta_2 - 1)^{-1}(1 - e^{-t/t_0})$$
$$t_0^{-1} = 2\nu(\eta_2 - 1). \tag{3.9}$$

The second term in expression (3.9) for $\bar{n}_0(t)$ takes into account the change in the initial number of photons due to spontaneous emission from atoms when $0 < t < t_0$. When $t > t_0$ the value of $\bar{n}_0(t)$ approaches the steady state value $\bar{n}_0 = (1-\eta_1)^{-1} + (\eta_2 - 1)^{-1}$. It can approximately be maintained that in the first stage we have a linear process of amplification of the thermal noise, in which the average number of photons initially is $\bar{n}_0$. In this stage the fluctuations of the photons are large, $\overline{\Delta n^2(t)} = \bar{n}^2(t)$, and saturation has no significance.

In the following stage when $t_0 < t < t_0 \ln (n_\infty/\bar{n}_0)$ (here n_∞ is the steady state value of the number of photons) the saturation effect must be taken into account. The exponential growth of the field stops, and the fluctuations begin to decrease. In this stage spontaneous emission from the atoms may be neglected, since the corresponding noise no longer has time to be amplified.

A strict description of the entire process requires the solution of the time-dependent Fokker–Planck equation for $\rho(t\xi)$. Approximately, however (excluding large values of the time $t > t_0 \ln (n_\infty/\bar{n}_0)$) both stages of the transition process can be described by solving the classical equations for the field $n(t)$ with the appropriate initial conditions

$$n(t) = \frac{n_0}{e^{-t/t_0} + n_0/n_\infty}, \qquad n_0 \ll n_\infty. \tag{3.10}$$

The initial value of the number of photons n_0 is distributed according to the Planck law with the average value $\bar{n}_0$. Hence we can find the distribution function of the photons $P_n(t)$ at time t:

$$P_n(t) = \frac{1}{\bar{n}_0}\left(1 - \frac{n}{n_\infty}\right)^2 \exp\left[-\frac{t}{t_0} - \frac{n}{\bar{n}_0}\left(1 - \frac{n}{n_\infty}\right)^{-1} e^{-t/t_0}\right]. \tag{3.11}$$

Figure 6a and b shows the results of calculations (the solid line) and measurements of the first two moments $\bar{n}(t)$ and $\overline{\Delta n^2(t)}$ [31]. As the steady state is approached, the fluctuations decrease and approach Poisson's law $\overline{\Delta n^2} \sim \bar{n}$. This value is much smaller than the fluctuations in the initial stage of amplification, where $\overline{\Delta n^2} \sim \bar{n}^2$. Thus, in the process of reaching a steady state the dispersion of the photons has an appreciable maximum.

It must be noted that when $t \gg t_0$ formula (3.11) is no longer applicable, since it does not take into account the fluctuations in the steady state.

Another method of observing the emission fluctuations in transient conditions is based on studying the distribution of triggering times of a laser [32, 33]. A certain level of oscillation n is set and the time t

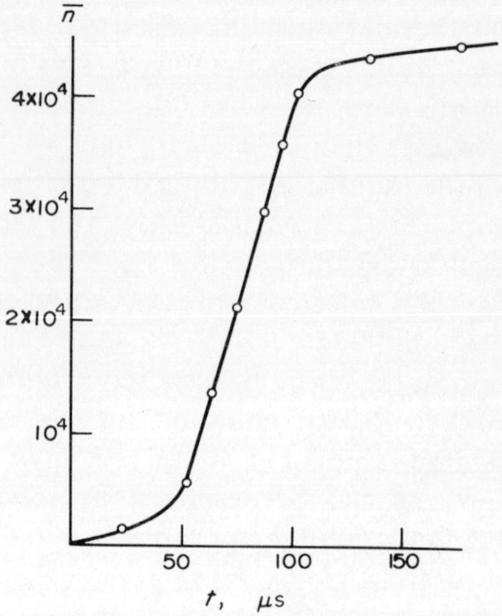

FIG. 6a. Time dependence of the average number of photons found in ref. 37 shown by solid line. This curve was used for adjusting the unknown parameters in the theory.

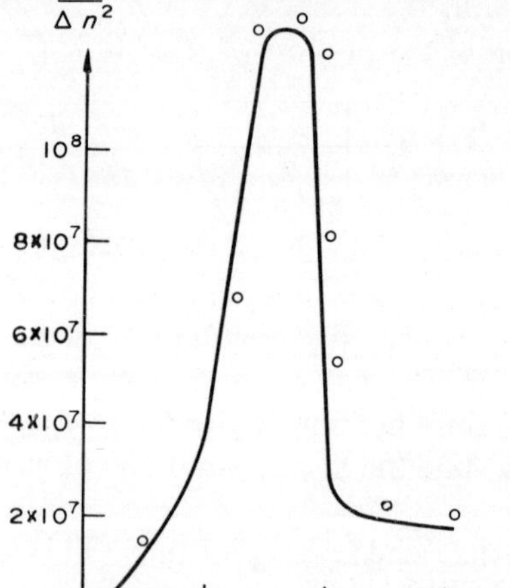

FIG. 6b. Dispersion of photons as a function of time. The solid line shows the theoretical results for the parameters $\bar{n}_0 \approx 2 \times 10^2$, $n_\infty \approx 4 \times 10^4$ and $t_0 \approx 10^{-5}$ s. The experimental results were obtained in ref. 31.

during which this intensity is reached is measured. This method can be used to find the distribution function $P_t(n)$ which determines the probability of triggering $P_t(n)\,\Delta t$ in the time interval $(t, t + \Delta t)$. The distribution function $P_t(n)$ is related to (3.11) by the expression

$$P_t(n)\,\Delta t = P_n(t)\,\Delta n.$$

Figure 6c shows ln $[P_t\Delta t]$ calculated with the aid of the distribution (3.11) (solid line) and found experimentally [33].

The statistical properties of radiation in transient conditions are

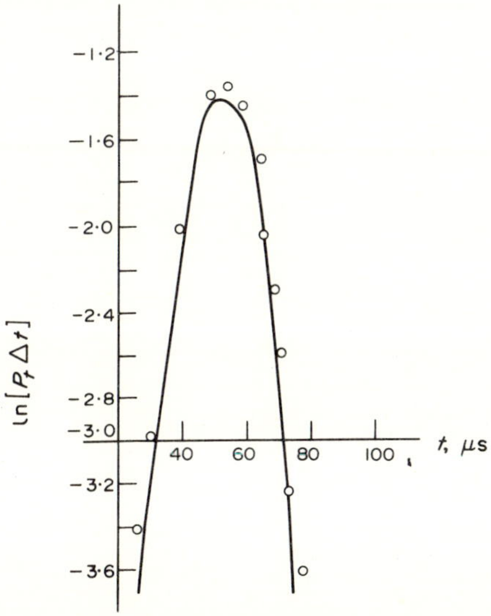

FIG. 6c. Value of ln $[P_t\Delta t]$ as a function of the moment of time of laser triggering. The solid line corresponds to the theoretical curve plotted for the parameters $n/\bar{n}_0 \approx 10^3$, $n_\infty/n \approx 50$ and $t \approx 6\times10^{-6}$ s. The experimental points were obtained in ref. 33.

thus determined to a considerable extent by the initial distribution of the photons. If the initial distribution is changed, for example by mixing thermal radiation with a coherent field, then the fluctuations in the transition conditions will be smaller. In particular, the scatter of the triggering times will be diminished. Such an effect was observed in ref. [34].

The theory of transient conditions in a gas laser was developed in refs. [35]–[37].

The linewidth of the radiation

Above the oscillation threshold the linewidth of radiation is expressed by the formula

$$\Delta\nu = \nu(1 + \eta)/4\bar{n}. \tag{3.12}$$

Near the threshold eq. (3.12) transforms into the Townes formula taking account of a factor of 2[2]: $\Delta\nu = \nu/2\bar{n}$. Far from threshold the linewidth reaches a limit that does not depend on the field energy $\Delta\nu = \nu\beta/4$. The linewidth was measured experimentally in ref. [38]–[41].

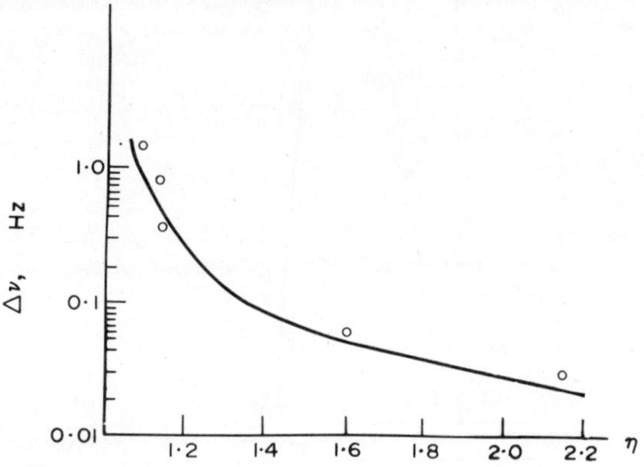

FIG. 7. Dependence of the radiation linewidth on η. At threshold we have $\Delta\nu \sim \nu\sqrt{\beta}$. When $\eta - 1 \gg \sqrt{\beta}$ the linewidth is described by formula (3.12) (solid line). The points correspond to measurements made in ref. 39.

The results of measurements performed in ref. [39] can be plotted with the curve $\Delta\nu(\eta)$ determined by eq. (3.12) and shown in Fig. 7.

The time correlation of photons

Above the oscillation threshold the correlation function of fluctuations of photons has the form

$$G(t) = \overline{\Delta n^2}\, e^{-\Gamma t}, \qquad \Gamma = 2\nu(\eta - 1). \tag{3.13}$$

When $|\eta - 1| \sim \sqrt{\beta}$ the quantity Γ has a value of the order of $2\nu\sqrt{\beta}$. Upon further reduction of the gain parameter when $1 - \eta > \sqrt{\beta}$ the damping decrement again grows. This relation is shown schematically in Fig. 8. Strictly speaking, near the threshold itself the correlation function $G(t)$ can be written in the form of a sum of exponents with

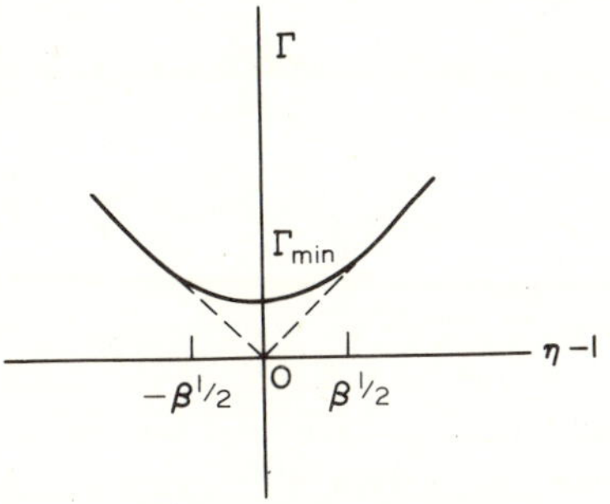

FIG. 8. Damping decrement of the intensity fluctuations Γ versus the generation parameter. When $|\eta - 1| > \sqrt{\beta}$ the decrement has the asymptotic values $\Gamma = 2\nu|\eta - 1|$. When $|\eta - 1| < \sqrt{\beta}$ the decrement is of the order of $\Gamma_{min} \sim 2\nu\sqrt{\beta}$.

different damping decrements [42, 43]. In other words, the frequency spectrum of fluctuations in this case differs somewhat from Lorentzian. This difference has been confirmed experimentally [24].

Very near the oscillation threshold the time correlation of the photons (including high order correlations) were studied in refs. [44]–[46] with the aid of photocounting techniques.

Above the oscillation threshold the frequency spectrum of intensity fluctuations was measured in refs. [47]–[49] by a heterodyne method.

3.2. The Coherent Interaction of Wave trains

Let us now consider case (B) (subsection 1.4). The time ratio characteristic of this case can be interpreted as follows: $c\tau$ is greater than the free path of a photon c/ν (c is the velocity of light). An electromagnetic train emitted by an atom immediately interacts with several other atoms. A certain correlation appears in the spontaneous emission from the atoms, and to take it into account the correlation function δr_{12} must be calculated.†

It should be noted that the criterion "$c\tau >$ photon path" also appears in the theory of the diffusion of resonance radiation in a gas; in this case capture of the radiation occurs. A photon emitted by one atom does not have time to escape but is absorbed by another atom. The energy is mainly concentrated in the excited atoms. The fraction of the energy in the photons has the order of magnitude of $1/\nu\tau$.

The criterion which governs the importance of this capture is usually formulated as the condition for the density of the excited atoms

$$\lambda\!\!\!^{-2} c\tau N/V \gtrsim 1. \tag{3.14}$$

Here $\lambda\!\!\!^{-2}$ is the cross-section of resonance absorption (emission) of a photon, where $2\pi\lambda\!\!\!^{-}$ is the wavelength of the radiation.

In a weak field $N \approx N_{th} = \nu/g^2\tau$. If we consider the linewidth to be the natural linewidth, we arrive at the criterion $\nu\tau \lesssim 1$. In a strong field the cross-section for the absorption of a photon decreases, due to saturation, approximately as $1/\eta$, and the number of atoms grows as $N = \eta N_{th}$. Consequently the criterion for capture $\nu\tau \lesssim 1$ does not depend on the energy of the field.

In a strong field the level of the oscillation becomes unstable. An analysis of the stability [50, 51] shows that stationary oscillations are stable only outside the hatched area in the plot of η versus $\nu\tau$ (Fig. 9, see also Appendix II). Inside the hatched area the amplitude of the output itself oscillates [52, 53]. Let us find a solution of the steady state quantum problem in the region of parameters where the output amplitude is constant.

Below the threshold and at the threshold the distribution of the photons does not appreciably differ from the case of a gas laser. Above the threshold we have the distribution function (2.14) and the following

† Collective effects in the spontaneous emission of atoms are quite appreciable in the theory of superradiance [56]. This problem is discussed in refs. [21], [62], [63] and in the review [81].

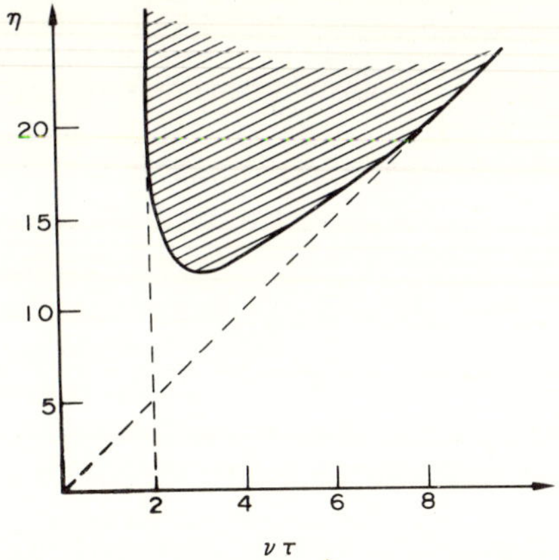

FIG. 9. The region of instability in the plot of η versus $\nu\tau$ is hatched. Inside this area the amplitude of oscillation varies in an oscillatory manner. Outside the hatched area the steady state oscillation amplitude is stable.

expression for the dispersion of the photons:

$$\overline{\Delta n^2}/\bar{n} = 1 + (\eta - 1)^{-1} \times \frac{2 + \nu\tau + 2\nu\tau\eta(\eta - 1)}{4\nu\tau + (\nu\tau)^2 + (2 - \nu\tau)\eta}. \tag{3.15}$$

Near the threshold $\overline{\Delta n^2}/\bar{n} \approx (\eta - 1)^{-1}(1 + \nu\tau)^{-1}$, i.e. the relative fluctuations due to collective effects decrease by the factor $(1 + \nu\tau)^{-1}$ in comparison with eq. (3.6).

The denominator in formula (3.15) becomes zero on the curve

$$\eta_c = \frac{\nu\tau(\nu\tau + 4)}{\nu\tau - 2} \tag{3.16}$$

which is the boundary of the unstable region.

Physically this can be interpreted as follows. Let us consider two dipoles at a distance from each other of the order of the free path of a photon. If one dipole emits a wave with the phase φ, then the second dipole in the case of exact resonance $\omega_0 = \omega_{ab}$ has a phase of oscillations $\varphi + \pi/2$, and the train emitted by it will have the phase $\varphi + \pi$. The scattered radiation will decrease the incident one. The resulting radiation will be weakened the more, the larger the overlap of the trains i.e. the larger the coherence parameter $\nu\tau$. It should be emphasized that here we are dealing with spontaneously emitted trains, since in induced radiation the trains emitted by different atoms have the same phases.

In a strong field E the energy of the atomic states splits and a detuning of the order of $2d_{ab}E$ appears between the atomic dipoles. When $d_{ab}E > (\nu/\tau)^{\frac{1}{2}}$ the scattering of the electromagnetic train by an atom has a nonresonant nature, so that the resulting phase shift is close to 0 or 2π. Thus the scattered radiation amplifies the incident one and the fluctuations may become anomalously great. It is easy to see that the condition $d_{ab}E \approx (\nu/\tau)^{\frac{1}{2}}$ coincides with the boundary of the region of instability $\eta_c \approx \nu\tau$ at larger energies.

Within a small area near the region of instability $\eta - \eta_c \sim 1/\sqrt{\bar{n}}$, formula (3.15) obtained from perturbation theory can no longer be applied.

The radiation linewidth in the region of stability is expressed by the relation

$$\Delta\nu = \frac{\nu(1+\eta)}{4\bar{n}(1+\nu\tau)} \qquad (3.17)$$

so that at large $\nu\tau$ the linewidth is determined by the lifetime of an atom $\Delta\nu = (1+\eta)/4\bar{n}\tau$.

Let us now consider the time correlation of intensities. Near threshold we have, as previously, formula (3.13) with the damping decrement $\Gamma = 2\nu(\eta - 1)(1 + \nu\tau)^{-1}$. When the energy grows, oscillations appear in the correlation function $G(t)$. Near the boundary of the region of instability, the correlation function is completely undamped [54]

$$G(t) = \overline{\Delta n_c^2} \cos \omega_c t$$

$$\omega_c^2 = \frac{\nu}{\tau}(\eta_c - 1)(1 + \nu\tau/2). \qquad (3.18)$$

In conclusion it should be noted that oscillations in the output of a single-mode CO_2-laser with nonlinear absorption was observed in ref. [55]. The coherence parameter $\nu\tau_{ab}$ in the absorbing cell was large, and operation was far above threshold. It is possible that such an oscillation regime is connected with getting into the unstable region shown in Fig. 9.

3.3. The Incoherent Interaction of Wave trains

Finally, let us consider case (C), subsection 1.4. Since $\tau_{ab} \ll \tau$, the spontaneous electromagnetic train $c\tau$ consists of short trains $c\tau_{ab}$ that are not correlated with each other in phase. For this reason when the inequalities $\nu\tau > 1$ and $\nu\tau_{ab} \ll 1$ hold we can also speak of the interaction of trains, but now it has an incoherent nature.

In this case the steady state oscillation is stable. However, at large values of τ the damping decrement of small perturbations of the order of τ^{-1} also becomes small. As a result the level of fluctuations may become high, i.e. of the order of $\nu\tau$. For the same reason oscillations appear in the intensity correlation function.

Below is an expression for the dispersion of photons above the generation threshold:

$$\overline{\Delta n^2}/\bar{n} = \frac{\eta}{\eta - 1} - \frac{1}{2\eta} + \nu\tau\frac{(\eta + 1)}{\eta}. \tag{3.19}$$

Here the first two terms on the right-hand side correspond to formula (3.7), and the last term appears because of collective effects.

The parameter $\nu\tau$ may be very large, of the order of 10^4 to 10^5 for ruby and CO_2 lasers. Hence the level of fluctuations even at a small amount above threshold is determined by the parameter $\nu\tau$. Far from threshold the relative fluctuations tend to a constant limit

$$\overline{\Delta n^2} = \bar{n}\nu\tau. \tag{3.20}$$

This relation can be interpreted as follows. A photon beam is divided into groups consisting of $\nu\tau$ photons correlated with each other. The separate groups of photons may be considered independently distributed according to Poisson's law.

The radiation linewidth in this case is determined by formula (3.12) for a gas laser, i.e. the effects of a self-consistent field in the case of incoherent summation of wave trains do not contribute to phase diffusion.

The intensity correlation function $G(t)$ is now expressed in the form of the sum of two exponents with decrements Γ_1 and Γ_2

$$\Gamma_{1,2} = \frac{\eta}{2\tau}\{1 \pm \sqrt{[1 - 8\nu\tau(\eta - 1)/\eta^2]}\}. \tag{3.21}$$

Here the following cases are possible at large values of $\nu\tau$. Near the threshold itself, when $\nu\tau(\eta - 1) \ll 1$, it is necessary to retain only the exponent with the smaller damping decrement $\Gamma \approx 2\nu(\eta - 1)$, so that we again return to case (3.13).

Further, when the inequality $8\nu\tau(\eta - 1) > 1$ is obeyed, the correlation function stops being monotonic. Oscillations appear in it having a frequency ω_c

$$\omega_c \approx \sqrt{[2\nu(\eta - 1)/\tau]}. \tag{3.22}$$

In this case a peak appears in the frequency spectrum of the intensity fluctuations at $\omega \approx \omega_c$ [57]. This frequency is in general large in comparison with the damping decrement $1/\tau$. Finally, very far above the threshold ($\eta > 8\nu\tau$), the oscillations vanish and $G(t)$ monotonically damps with time.

The case considered above was investigated theoretically in refs. [58]–[61], and the peak in the frequency spectrum and the square root dependence (3.22) on the excess above threshold predicted by McCumber were experimentally observed in ref. [64].

4. SOME APPLICATIONS OF THE THEORY

In the present section we consider fluctuations in two special types of lasers, namely, in a laser with nonlinear absorption and in a laser with a nonmonotonic dispersion contour. These examples are of interest from the point of view that the fluctuations in them may be anomalously great. In the first case we deal with intensity fluctuations, and in the second with phase fluctuations.

4.1. A Laser with Nonlinear Absorption

A gas laser with nonlinear absorption was studied experimentally in refs. [65], [66]. In these experiments a neon absorption cell was introduced into the Fabry–Perot resonator together with the active He–Ne. When the absorption in the passive cell is increased the plot of the radiated power behaves approximately as shown in Fig. 10.

A laser with nonlinear absorption is described by two gain parameters

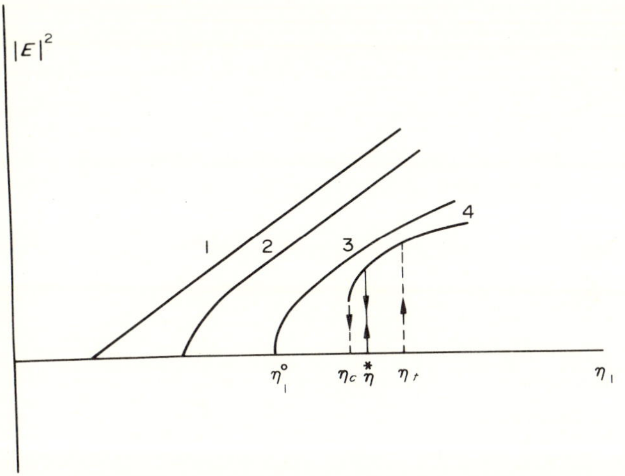

FIG. 10. Dependence of the radiated power on the gain parameter η_1 of the active cell. Straight line 1 corresponds to an ordinary laser with $\eta_2 = 0$. Curve 2 is for a small, but finite absorption $0 < \eta_2 < \eta_2^0$. Curve 3 applies to critical absorption and is described by formula (4.1). When $\eta_2 > \eta_2^0$ hysteresis occurs: with a growth of η_1 triggering occurs at point η_t, and with a reduction of η_1 cutoff occurs at point η_c. For a very slow change in η_1 triggering and cutoff occur at $\overset{*}{\eta}$.

η_1 and η_2 (η_1 is proportional to the number of atoms in the active cell, and η_2 to the number of atoms in the passive cell). We have straight line 1 for an ordinary laser, without nonlinear absorption, $\eta_2 = 0$. At a low but finite value of η_2 the plot of the radiated power deviates from a straight line (curve 2).

Further, at a certain critical absorption $\eta_2 = \eta_2^0$ the power plot has the form (curve 3):

$$|E|^2 \sim \sqrt{(\eta_1 - \eta_1^0)}, \qquad \eta_1 \geqslant \eta_1^0. \tag{4.1}$$

A further growth of absorption is accompanied by the appearance of hysteresis. As η_1 is increased triggering occurs at $\eta_1 = \eta_t$, and as η_1 is reduced oscillation stops at $\eta_1 = \eta_c$.

The picture shown in Fig. 10 corresponds to the liquid–gas phase transition near the critical point. Here the gain parameter η_1 corresponds to the pressure, and the radiation energy to the volume of the system, so that curves 1–4 are isotherms. The parameter η_2 plays the part of the temperature.

The state near the triggering point where oscillation does not occur and the oscillatory state near the cutoff point are both metastable. They are similar to superheated and supercooled phases. Thus curve 3 for $\eta_2 = \eta_2^0$, which separates the region without hysteresis, is equivalent to a critical isotherm, and the point $\eta_1 = \eta_1^0, \eta_2 = \eta_2^0$ is a critical point. Hysteresis phenomena are also observed when the frequency is varied.

The classical theory of the gas laser with nonlinear absorption was considered in refs. [67], [68], [84].

Let us introduce the total gain coefficient

$$\begin{aligned} \chi(\xi) &= \chi_1(\xi) - \chi_2(\xi) - \nu \\ \chi_i(\xi) &= \nu\eta_i(1 - \beta_i\xi + \beta_i^2\xi^2), \qquad (i = 1, 2). \end{aligned} \tag{4.2}$$

Here the subscript 1 relates to an active cell and 2 to a passive one. Since the most noticeable effects in the fluctuations occur near the critical point, we have limited ourselves to an expansion with respect to the field. The expansion should be carried out to the second order, since at the critical point the constant coefficient and the factor of ξ become equal to zero. From this condition we find the critical values η_1^0 and η_2^0:

$$\eta_1^0 = \frac{\beta_2}{\beta_2 - \beta_1}, \qquad \eta_2^0 = \frac{\beta_1}{\beta_2 - \beta_1}. \tag{4.3}$$

Hence it follows that hysteresis is possible only when

$$\beta_2 > \beta_1. \tag{4.4}$$

The meaning of this condition is very simple. The role of the passive cell is to make the gain coefficient a nonmonotonic function of the field energy. This can be achieved only if the absorption coefficient in the passive cell changes sufficiently rapidly, i.e. the saturation parameter β_2 is sufficiently large.

Near the critical point, when $\delta\eta_i = \eta_i - \eta_i^0 \ll \eta_i^0$ the total gain coefficient has the form

$$\chi(\xi) = \nu(\delta\eta - \beta_0\xi - \beta^2\xi^2)$$
$$\beta^2 = \beta_1\beta_2; \qquad \beta_0 = \beta_2\delta\eta_2 - \beta_1\delta\eta_1; \qquad \delta\eta = \delta\eta_1 - \delta\eta_2. \tag{4.5}$$

Let us now consider certain features of the fluctuations near the critical point [69]. The stationary photon distribution function $\rho_0(\xi)$ corresponding to the gain coefficient (4.5) on curve 3 ($\delta\eta_2 = 0$) is equal to

$$\rho_0(\xi) = \text{const.} \exp(\delta\eta\xi - \beta^2\xi^3/3). \tag{4.6}$$

Here for simplification of the formulae we have omitted the unimportant factor $1/\eta_1^0$ in the exponent. The average number of photons and the dispersion found with the aid of this function are shown schematically in Fig. 11. A glance at this figure shows that there are two

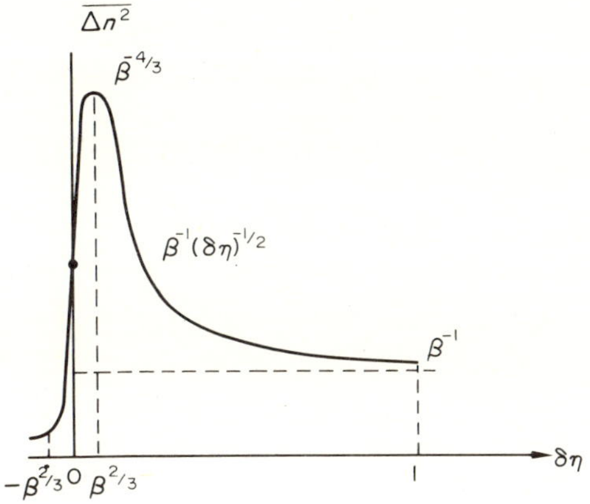

FIG. 11. Dispersion of photons as a function of the generation parameter $\delta\eta$ for critical absorption $\eta_2 = \eta_2^0$. When the threshold is passed in the small region $|\delta\eta| \sim \beta^{2/3}$ the dispersion sharply grows to the value $\Delta n^2_{max} = \beta^{-4/3}$. Upon a further growth of $\delta\eta$ the dispersion decreases as $(\delta\eta)^{-1/2}$ and at $\delta\eta \sim 1$ becomes a quantity of the order β^{-1} characteristic of an ordinary laser. Thus, Δn^2_{max} exceeds the dispersion of photons in an ordinary laser $\beta^{-1/3}$ times. In a gas laser $\beta^{-1/3} \sim 10^2$ to 10^3.

characteristic regions: a narrow transition region $|\delta\eta| \sim \beta^{2/3}$ and a region $\delta\eta \sim 1$ in which the functions change comparatively slowly. We continued the function $\overline{\Delta n^2}(\delta\eta)$ up to values of $\delta\eta \sim 1$, where this function is of the same magnitude as in an ordinary laser. The transition from thermal radiation to classical oscillation now takes place in a narrower region $|\delta\eta| \sim \beta^{2/3}$. It should be recalled that in an ordinary laser the width of the transition region is of the order $\beta^{1/2}$.

The dispersion of photons at the maximum is a quantity of the order of $\beta^{-4/3}$, which is $\beta^{-1/3}$ times (i.e. two or three orders) larger than in a laser without nonlinear absorption. The decrease in dispersion with an increase in the gain parameter follows the law $\overline{\Delta n^2} \sim \beta^{-1}(\delta\eta)^{-1/2}$.

Figure 12 shows the dependence of the damping decrement of the

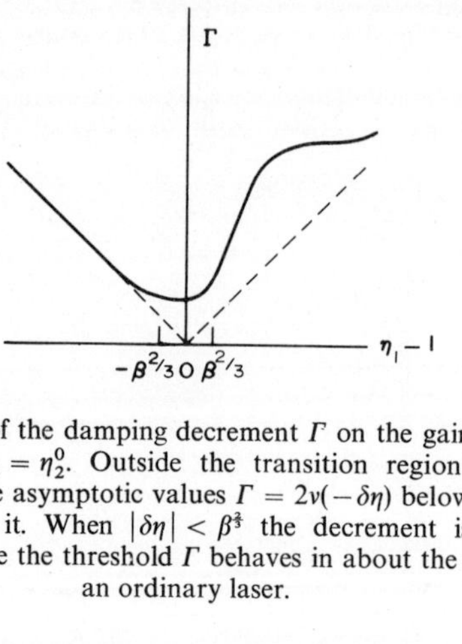

FIG. 12. Dependence of the damping decrement Γ on the gain parameter for critical absorption $\eta_2 = \eta_2^0$. Outside the transition region $|\delta\eta| \sim \beta^{2/3}$ the decrement acquires the asymptotic values $\Gamma = 2\nu(-\delta\eta)$ below the threshold, and $\Gamma = 4\nu\delta\eta$ above it. When $|\delta\eta| < \beta^{2/3}$ the decrement is of the order $\Gamma_{min} \sim 2\nu\beta^{2/3}$. Far above the threshold Γ behaves in about the same way as in an ordinary laser.

intensity correlation function Γ on $\delta\eta$ near the critical point. This dependence differs from the one given in Fig. 8 both at, and above, the oscillation threshold. At the threshold, instead of $\Gamma_{min} \sim 2\nu\beta^{1/2}$ in an ordinary laser, we now have $\Gamma_{min} \sim 2\nu\beta^{2/3}$. Above the threshold we have $\Gamma = 4\nu\delta\eta$, i.e. here the damping decrement is twice as large. The dotted line shows the asymptotic form $2\nu\delta\eta$, departure from which occurs at $\delta\eta \sim 1$.

Thus the main features of the fluctuations near the critical point are that the transition threshold region sharply narrows, and the dispersion of photons greatly grows in this region. An appreciable maximum appears in the dependence of the dispersion of the photons on the generation parameter.

Let us now consider the region of hysteresis, i.e. we consider the case $\delta\eta_2 > 0$. We shall assume that the excess above the critical value is small, $\theta = \delta\eta_2/\eta_2^0 \ll 1$.

In this case the triggering and cutoff points are close to the critical point, $\eta_c - \eta_1^0 \sim \theta$ and $\eta_t - \eta_1^0 \sim \theta$. Here $\eta_t - \eta_c \sim \theta^2$. At the triggering and cutoff points $\bar{n} \sim \beta^{-1}\theta$.

It is clear that at very low values of θ the fluctuations blur the hysteresis picture. Hysteresis can be observed only at $\theta \gtrsim \beta^{\frac{1}{3}}$. This can easily be obtained from the condition $\overline{\Delta n^2} \lesssim \bar{n}^2$, or from the condition that $\eta_t - \eta_c$ is larger than the width of the transition region at the critical point $\beta^{\frac{2}{3}}$.

In the hysteresis region the fluctuations may become larger near the cutoff point (curve 4, Fig. 10). When approaching the cutoff point, the dispersion of the photons has a characteristic square root dependence:

$$\overline{\Delta n^2} = \beta^{-1}(\eta_1 - \eta_c)^{-\frac{1}{2}}. \tag{4.7}$$

At the cutoff point itself it should be assumed that $\eta_1 - \eta_c \sim \beta^{\frac{2}{3}}$, so that $\overline{\Delta n^2_{\max}} \sim \beta^{-\frac{4}{3}}$. The fluctuations at the cutoff point have approximately the same magnitude as at the critical point. The damping decrement accordingly has the form

$$\Gamma = 2\nu(\eta_1 - \eta_c)^{\frac{1}{2}}\,\theta. \tag{4.8}$$

At the cutoff point $\Gamma_c \sim 2\nu\beta^{\frac{1}{3}}\theta$.

Let us now see what the form of the photon distribution function is in the hysteresis region. In this region $\rho_0(\xi)$ has two maxima at $\xi = 0$ and $\xi = \bar{n}$, which determine the probability of states without oscillation and with oscillation.

As η_1 approaches the triggering point, $\rho(\bar{n})$ grows and $\rho(0)$ decreases. Conversely, when approaching the oscillation cutoff point we have the relation $\rho(0) \gg \rho(\bar{n})$. The more probable state corresponds to one in thermodynamic equilibrium, and the less probable state to a metastable one. Since the lifetime of the metastable state is exponentially large, it is worthwhile to speak of the statistical properties of this state (of fluctuations of the photons and the radiation linewidth).

Let us estimate the lifetime of the metastable state. When η_1 decreases and approaches the value η_c, the probability of oscillation cutoff grows appreciably. For the lifetime of the state with oscillation we approximately have

$$T_c \sim \nu^{-1} \exp\left[\tfrac{4}{3}(\eta_1 - \eta_c)^{\frac{3}{2}} \beta^{-1}\right], \qquad \eta_1 > \eta_c \tag{4.9}$$

A similar estimate of the time of spontaneous triggering with a growth of η_1 and when approaching η_t gives

$$T_t \sim \nu^{-1} \exp\left[\tfrac{1}{4}(\eta_t - \eta_1)^{\frac{3}{2}} \beta^{-1}\right], \qquad \eta_t > \eta_1. \tag{4.10}$$

Near the cutoff point $T_c \ll T_t$ and, conversely, near the triggering point $T_t \ll T_c$. At a certain value of $\eta_1 = \overset{*}{\eta}$ we have $T_c(\overset{*}{\eta}) = T_t(\overset{*}{\eta})$, whence we find

$$\overset{*}{\eta} = (3\eta_c + \eta_t)/4. \tag{4.11}$$

When $\eta_1 = \overset{*}{\eta}$ the states with and without oscillation become equally probable. For this reason, with a very slow change in the generation parameter (during a time of the order $T(\overset{*}{\eta})$), the hysteresis disappears; oscillation sets in and vanishes at the point $\eta_1 = \overset{*}{\eta}$. This transition is shown by a solid line in Fig. 10.

Thus, when the gain parameter is slowly changed we are dealing with a phase transition of the first kind when $\eta_1 = \overset{*}{\eta}$. The method of determining $\overset{*}{\eta}$ considered above completely conforms with the well-known Maxwell phase rule.

If we fix the gain parameter at a value of $\overset{*}{\eta}$, then oscillation will take place in a pulsed regime. The pulsations have a random nature.

Let us estimate the values of the parameter at which it would be possible to observe such pulsations. To obtain a pulsation period of less than one second, the exponent in eqs (4.9) and (4.10) should not exceed 15. Hence we estimate the permitted range of gain parameters $\delta\eta_1 \gtrsim 15\beta^{\frac{2}{3}}$, i.e. this interval is about 15 times greater than the width of the transition region at the critical point. The gain parameter in a passive cell must be fixed with the same accuracy. For $\beta \sim 10^{-6}$ to 10^{-8} we have $\delta\eta_1 \sim 10^{-3}$ to 10^{-4}.

Scattering in the time of switching on and off of a laser with nonlinear absorption was studied in ref. [70]. Scattering of the switching-off time was observed when approaching the oscillation cutoff point, which is connected with a growth in the fluctuations at this point.

Large intensity fluctuations in a laser with nonlinear absorption are connected with the special behaviour of $\chi(\xi)$ near the oscillation threshold. The dispersion of photons and the damping decrement Γ in the region of classical oscillation are related to the derivative of the gain coefficient

$$\overline{\Delta n^2} = \nu \Big/ \left(\frac{\partial \chi}{\partial \xi}\right)_{\xi = \bar{n}}; \qquad \Gamma = 2\bar{n}\left(\frac{\partial \chi}{\partial \xi}\right)_{\xi = \bar{n}}. \tag{4.12}$$

Near the critical point and the cutoff point $(\partial\chi/\partial\xi)_{\xi=\bar{n}}$ is small and the fluctuations grow.

4.2. A Laser with Anomalously Large Phase Fluctuations

Do conditions exist in which the frequency fluctuations become large? We shall show by using a simple model that such conditions exist and they are connected with features in the behaviour of the imaginary part χ'' of the susceptibility of the medium [71].

Let us consider a gas laser with an absorbing cell and having the following parameters: (1) the frequencies of the atomic resonances of the active and passive cells coincide, (2) the Doppler contour in the passive cell is narrower than in the active one, $v_1 > v_2$ (v_i is the mean thermal velocity of the atoms in the ith cell, the subscript 1 relates to the active cell and 2 to the passive one), (3) the saturation parameter in the passive cell is less than in an active one, $\beta_2 \leqslant \beta_1$.

The last condition has been introduced only to simplify the situation by the following consideration: it allows us to assume that the gain coefficient is monotonic with changes in the field. For this reason the nonlinear effects considered in subsection 4.1 are absent in this case. For the same reason, considering the field to be weak, we shall not take it into account when considering frequency effects. The total susceptibility $\chi = \chi' + i\chi'' = \chi_1 - \chi_2 - \nu$ becomes complex when we consider finite detunings of the frequency $\Delta\omega = \omega - \omega_0$. Let us find χ' (the gain coefficient) and χ'' for the case when there is large Doppler broadening in both cells, $kv_i \gg 1/\tau_i$.

Considering $\Delta\omega$ to be small, we shall not take into account the dependence of the gain coefficient on detuning:

$$\chi' = \nu[\eta_1 - \eta_2 - 1 - \xi(\eta_1\beta_1 - \eta_2\beta_2)] \tag{4.13}$$

$$\eta_i = \sqrt{(\pi)}\, g^2 N_i/kv_i\nu \qquad (i = 1, 2)$$

In the expansion of χ'' with respect to $\Delta\omega/kv_i$ we shall not take into account the dependence of the coefficients a and b on the field, since $\beta_i\xi \ll 1$

$$\chi'' = a\Delta\omega + b(\Delta\omega)^3 \tag{4.14}$$

$$a = \eta_1\varepsilon_1 - \eta_2\varepsilon_2; \qquad b = \frac{\pi}{6\nu^2}(\eta_2\varepsilon_2^3 - \eta_1\varepsilon_1^3)$$

$$\varepsilon_i = 2\nu/\sqrt{(\pi)}\, kv_i.$$

The radiation linewidth is proportional to $(1 + a)^{-2}$. In a gas laser the coefficient $1 + a$ is usually close to unity. We, however, shall be interested in the case when $1 + a$ becomes equal to zero. This condition is similar to the definition of the critical point (4.3).

Let us find the critical values of the parameters η_1^0 and η_2^0 at which the conditions $\chi' = 0$ and $1 + a = 0$ are complied with simultaneously. In the approximation where the radiation field vanishes we have

$$\eta_1^0 = \frac{1 + \varepsilon_2}{\varepsilon_2 - \varepsilon_1}; \qquad \eta_2^0 = \frac{1 + \varepsilon_1}{\varepsilon_2 - \varepsilon_1}. \tag{4.15}$$

It should be remembered that according to our assumption $\varepsilon_2 > \varepsilon_1$. For small values of ε_i both parameters η_1^0 and η_2^0 are large and approximately equal.

Let us estimate the order of magnitude of the radiation linewidth when $1 + a = 0$ with a small but finite radiation energy. For this purpose we can use eq. (1.15), since we are interested only in a slow fluctuation of the frequency, and the system of atoms has time to follow the field. Introducing $E(t) = |E| \exp(i \int dt\, \Delta\omega(t))$ into eq. (1.15), we get

$$\Delta\omega(t) = kv(\mathrm{Im}\, f(t)/\nu\,|E|)^{\frac{1}{3}} \tag{4.16}$$

where v is a certain average velocity between v_1 and v_2. In an ordinary laser a frequency fluctuation of the order of $\mathrm{Im}\, f(t)/\nu\,|E|$ is a very small quantity. According to eq. (4.16), the frequency fluctuations at the critical point are considerably greater.

To find the radiation linewidth $\Delta\nu$, let us compute the frequency correlation function

$$\Delta\nu = \int_0^\infty dt\langle\Delta\omega(t)\,\Delta\omega(0)\rangle. \tag{4.17}$$

We shall use the approximation $\langle \overset{*}{f}{}^{\frac{1}{3}}_{(t)} f^{\frac{1}{3}}_{(0)}\rangle \approx \langle \overset{*}{f}(t)\, f(0)\rangle^{\frac{1}{3}}$ in calculating this correlation function. We can take expression (1.15) as the correlation of the random force after first substituting a function of the form $1/\tau\, e^{-t/\tau}$, where τ is a certain average relaxation time ($\tau \sim \tau_1, \tau_2$) for the δ-function in time.

We finally have the following estimate for the radiation linewidth:

$$\Delta\nu \sim (kv)^2\,(\tau^2/\nu\bar{n})^{\frac{1}{3}} \tag{4.18}$$

A feature of this expression is that it includes the factor $\bar{n}^{-\frac{1}{3}}$ instead of $\bar{n}^{-1}$ as in an ordinary laser. If we assume that $kv \sim 1/\tau \sim 10^8$ Hz, $\nu \sim 10^7$ Hz and $\bar{n} \sim 10^8$, than $\Delta\nu \sim 10^6$ Hz. It is possible that conditions close to these can be realized in a He–Ne laser at 3.39 μm. The low value of the parameter ν/kv can be compensated for by operating well above threshold.

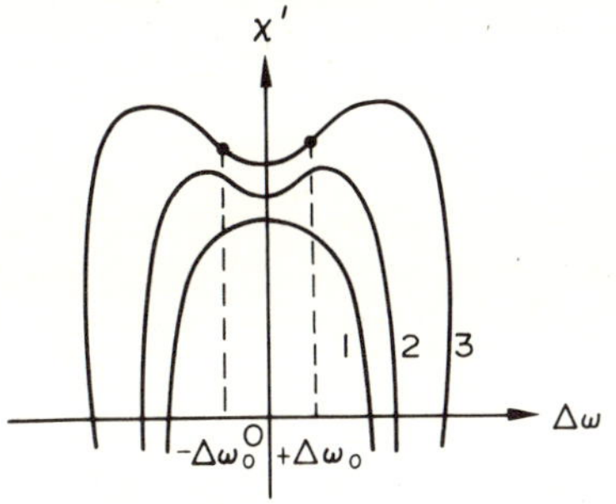

FIG. 13. Unsaturated gain coefficient χ' as a function of the frequency detuning $\Delta\omega$. Curve 1 corresponds to an ordinary laser oscillating at the centre of the line $\Delta\omega = 0$. Curve 2 is for critical absorption $\eta_2 = \eta_2^0$. Here oscillation also occurs at the centre of the line. Curve 3 corresponds to absorption larger than the critical value $\eta_2 > \eta_2^0$. The frequency of oscillation shifts from the centre of the line, so that the detuning is equal either to $+\Delta\omega_0$ or to $-\Delta\omega_0$.

Let us consider qualitatively the behaviour of the frequency and intensity when there is a further change in the parameters to the supercritical region. Plots of $\chi'(\Delta\omega)$ and $\Delta\omega + \chi''(\Delta\omega)$ are shown in Figs. 13 and 14. Curves 1, 2 and 3 correspond to the values $1 + a > 0$, $1 + a = 0$

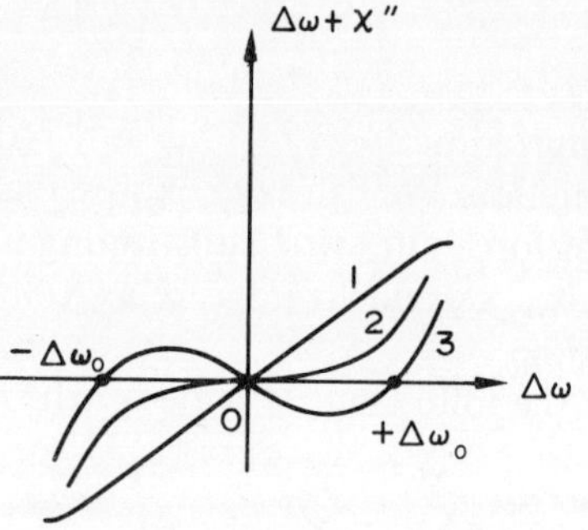

FIG. 14. Dispersion contour $\Delta\omega + \chi''(\Delta\omega)$. In an ordinary laser the dispersion contour is monotonic near the centre of the radiation line and approximately coincides with a straight line (line 1). For critical absorption the dispersion contour has the form of a cubic parabola const $(\Delta\omega)^3$ (curve 2). Curve 3 corresponds to the case $\eta_2 > \eta_2^0$. The dispersion equation has three roots, namely, $\Delta\omega = 0$, $\Delta\omega = \pm\Delta\omega_0$. In this case oscillation at the centre of the line is not stable.

and $1 + a < 0$ respectively. When $1 + a < 0$ the dispersion equation $\Delta\omega + \chi''(\Delta\omega) = 0$ has three solutions, namely, $\Delta\omega = 0$ and $\Delta\omega = \pm\Delta\omega_0$, where $\Delta\omega_0 = \sqrt{(-a/b)}$ and $a < 0$, $b > 0$.

Thus, when $1 + a < 0$ oscillation may occur at one of the three frequencies 0 and $\pm\Delta\omega_0$. However, oscillation at the frequency $\Delta\omega = 0$ is not stable. States with oscillation at the frequencies $\Delta\omega = \pm\Delta\omega_0$ have a large statistical weight, since the radiation energy in these states, as can be seen from Fig. 13, is greater than in the state with $\Delta\omega = 0$. The assumption that the field was weak was of importance here. In a strong field the picture may change qualitatively.

A large increase of the frequency fluctuations occurs only in a small region near the critical point (4.15). Let us estimate the magnitude of this region $\delta\eta_i = \eta_i - \eta_i^0$. Comparing the first and second terms in the expansion of (4.14), we find that

$$\delta\eta_1 \sim \delta\eta_2 \sim (v\tau\bar{n})^{-\frac{1}{3}}. \tag{4.19}$$

For the numerical example considered above $\delta\eta_i \sim 10^{-2}$.

In conclusion it should be noted that in the given case we are dealing with a phase transition of the second kind with respect to a shift in frequency $\Delta\omega$.† When $\eta_2 < \eta_2^0$ we have $\Delta\omega = 0$, and when $\eta_2 > \eta_2^0$ we have $\Delta\omega \sim \sqrt{(\eta_2 - \eta_2^0)}$. The equation $\Delta\omega + \chi''(\Delta\omega) = 0$ corresponds to the Ginzburg–Landau equation for determining the order parameter.

4.3. Shot Noise

Fluctuations appearing because of the discrete excitation mechanism of the atoms may sometimes compete with quantum fluctuations in a gas laser. Such fluctuations are similar to the shot noise in a vacuum tube oscillator.

The atoms in a gas laser are excited by collisions with atoms or electrons during a time that is small relative to the lifetime of an atom. For this reason the total radiating current is a superposition of the decaying currents of the separate atoms overlapping considerably in time. In these atoms the excitation times t_k are distributed randomly, and only the average rate of excitation of the atoms $\overset{\circ}{N} = N/\tau$ is given.

$$j = \sum_k gp(t - t_k) \tag{4.20}$$

where $p = \langle\sigma r\rangle$ is the average radiating current of an atom.

The shot noise is statistically independent of the quantum fluctuations, and for this reason it can be calculated separately.

Shot noise leads to a change in the diffusion coefficients B and C in the Fokker–Planck equation (3.4). The coefficients B' and C' connected with shot noise have the form (Appendix III)

$$B' = \xi(\eta - 1)/4\eta; \qquad C' = \xi(\eta - 1)(\Delta\omega\tau)^2/16\eta. \tag{4.21}$$

† Spencer and Lamb [82] have recently considered a laser model with two weakly coupled resonators. Here the dispersion contour is nonmonotonic too, but their approach is classical and neglects fluctuations.

These formulae relate only to the simplest model of a laser with homogeneous broadening. Comparing expressions (3.4) and (4.21), we see that shot noise is of importance when operating sufficiently far above threshold. This question is considered in greater detail in Appendix III.

An additional source of shot noise is the motion of an atom in the field of a standing wave [72]. It is well known that a moving atom emits a set of frequencies that are multiples of $2kv$. The radiation at multiple frequencies contains a random phase. For this reason such radiation gives an additional contribution to the incoherent component of the field in comparison with the case of a travelling wave.

4.4. Certain Other Questions

Up to now we have dealt only with the simplest model of a laser. The atoms were considered to be motionless and all of them were in identical conditions. It is quite obvious that such a model is too simplified. In a gas laser, for example, account must be taken of the thermal motion of the atoms and the spatial structure of the field. In addition, we assumed that only the upper laser level is excited. In real conditions in the absence of a field there is a population of the upper level N_a and of the lower level N_b, so that $\Delta N = N_a - N_b$ is the inversion. In a weak field only excited atoms contribute to spontaneous emission.

The intensity of the fluctuations is proportional to N_a. If we take this circumstance into consideration, then an additional factor $N_a/\Delta N$ appears in the expressions for the dispersion of the photons and the radiation linewidth. Thus, for example, above the threshold when $\eta - 1 \ll 1$ we have

$$\overline{\Delta n^2} = \frac{\bar{n}}{\eta - 1} \frac{N_a}{\Delta N}$$
$$\Delta \nu = \frac{\nu}{2\bar{n}} \frac{N_a}{\Delta N}. \tag{4.22}$$

In a gas laser the parameter $N_a/\Delta N$ varies within wide limits, $N_a/\Delta N \sim 1 - 10^2$.

It should be stressed, however, that the results obtained with the aid of a simple model are universal for a gas laser near the oscillation threshold. This means that the relations (4.22) do not depend on a particular model of a gas laser and are applicable for cases of homogeneous and inhomogeneous broadening, an arbitrary field model, etc. Their universal nature is due to the fact that a gas laser is described by the Fokker–Planck equation, and near the oscillation threshold the diffusion coefficient of photons in Eq. (1.18) is determined from thermodynamic relations. Far from threshold, where the system deviates considerably from equilibrium, such a universal nature disappears. For example, in the case of inhomogeneous broadening the fluctuations are different for travelling and standing waves [72]. It should be noted that in the case of coherent superposition of wave trains (subsection 3.2) the distribution function $\rho(t\xi)$ is not described by the Fokker–Planck equation. Here the level of the fluctuations depends on the details of the model even near threshold. For instance, in the case of homogeneous broadening according to formula (3.15), at threshold we have

$$\overline{\Delta n^2}/\bar{n} = (\eta - 1)^{-1}(1 + v\tau)^{-1},$$

and in the case of inhomogeneous broadening $\overline{\Delta n^2}/\bar{n} = (\eta - 1)^{-1}$ [54].

Another important generalization of a simple laser model takes into account several modes of the electromagnetic field in the resonator. These may be different axial modes, or modes differing in their direction of propagation (as in a ring laser), or modes with different polarizations. For two modes having field amplitudes z_1 and z_2 the weight function $\rho(z_1 z_2)$ determines the joint distribution of the fields z_1 and z_2.

Investigation of the cases when the modes interact with each other significantly is of the greatest interest. In steady state conditions, several stationary stable points of oscillation exist, generally speaking, on the plane ξ_1, ξ_2 in the classical region. Near each stable point the function $\rho(\xi_1 \xi_2)$ has a δ-like form. When one or several states become metastable and approach the region of instability, however, the distribution function at these points blurs. Here overlapping of the tails of the distribution functions corresponding to different stationary points may become noticeable. The degree of this overlap determines the lifetime of the metastable state; this is similar to the one-dimensional case in a laser with nonlinear absorption.

The two-mode distribution function $\rho(\xi_1 \xi_2)$ for different types of lasers was studied in refs. [73]–[76]. In some cases (travelling waves in a ring laser, the state of polarization of radiation at the transition $0 \rightleftarrows 1$), a laser may be in a state of indifferent equilibrium along a certain direction in the ξ_1, ξ_2 plane. If, for example, only the total radiation energy ξ has been fixed, then the straight line $\xi = \xi_1 + \xi_2$ is the position of indifferent equilibrium. Along this direction $\rho(\xi_1 \xi_2) = \text{const.}$, so that the fluctuations of ξ_1 and ξ_2 become large, and anticorrelation of the intensity fluctuations $\overline{\Delta n_1 \Delta n_2} = -\overline{\Delta n_2^2}$ occurs. Here account must be taken of factors that lead to the removal of degeneracy (for example, anisotropy of the mirrors, depolarizing collisions, etc.)

It should be noted that the effect of anticorrelation of the intensities in a two-mode laser was observed in ref. [77]. The spectrum of amplitude and phase fluctuations of radiation in two-mode conditions was investigated in ref. [78].

The special case when the condition for oscillation was satisfied for only one mode of the field, and remaining ones are close to the threshold, was considered in ref. [79].

In conclusion it should be noted that the two-mode conditions of oscillation may be used for measuring the quantum radiation linewidth [80].

5. CONCLUSION

The statistical properties of single-mode laser radiation are determined by the spontaneous emission from atoms of the active medium into the selected mode of the resonator. The spontaneous emission from atoms depends on a number of parameters of the medium and the resonator. The most important of these are the relaxation times of the atoms and the field. Depending on the ratio of these times, lasers may be divided into three groups (subsection 1.4).

In describing the fluctuations of laser radiation, difficulties appear connected with the fact that a laser is a nonequilibrium system. A rigorous method of describing such a system consists in solving quantum problems for the field and the medium. Using the natural small parameter in the quasiclassical region $1/\sqrt{\bar{n}}$, we can obtain a closed set of quantum kinetic equations. This system of equations can be used to describe any type of laser with an arbitrary energy.

It should be noted that the vicinity of the oscillation threshold can also be described by the phenomenological theory with random forces. This is connected with the circumstance that near the threshold we are dealing with a phase transition of the second kind, and here we can use thermodynamic relations. The quantum approach gives the same results.

The solution of the quantum problem makes it possible to find and describe the properties of radiation quite far above the threshold, when it is necessary to take into account the dependence of the probability of spontaneous emission of a photon on the energy of the field, and when collective effects in spontaneous emission are appreciable.

Fluctuations of radiation in a gas laser near the oscillation threshold have been studied experimentally in great detail. High moments of the distribution and the time correlations of photons have been measured by photocounting techniques. In the threshold region the relative fluctuations are large, of the order of unity. It is of interest to study other cases when the fluctuations may be large.

The dispersion of photons is inversely proportional to $(\partial\chi'/\partial\xi)_{\xi=\bar{n}}$, where χ' is the gain coefficient of the medium. Consequently the fluctuations become anomalously great where $(\partial\chi'/\partial\xi)_{\xi=\bar{n}}$ becomes equal to zero. This occurs in a gas laser with nonlinear absorption near the critical point and at the point of oscillation cutoff.

The quantum radiation linewidth is proportional to $(1 + \partial\chi''/\partial\Delta\omega)^{-2}$; where χ'' is the imaginary part of the polarizability of the medium. When $\partial\chi''/\partial\Delta\omega \to -1$, the frequency fluctuations grow quite appreciably. Such a case is realized in a laser with a nonmonotonic dispersion contour.

In these cases the fluctuations have an appreciable maximum near the critical points of the susceptibility $\chi(\xi\Delta\omega)$.

Another example of large fluctuations in the classical region of oscillation is the vicinity of the region of instability of single-mode oscillation (Fig. 9).

The experimental investigation of the fluctuations of laser radiation in the critical regions is undoubtedly of considerable interest.

APPENDIX I

Here we give an expression for the dispersion of photons and the radiation linewidth for the case of finite detuning $(\omega_{ab} - \omega_0)\tau = \varepsilon$ and finite population of the laser levels (N_a in the upper level, N_b in the lower one, $\Delta N = N_a - N_b$). Limiting ourselves to the region of oscillation above threshold, we have the following formulas for a gas laser [11]

$$\overline{\Delta n^2} = (1 + d)\bar{n}$$

$$d = \frac{N_a(\eta^{-1} + \varepsilon^2) + (\eta - 1)(1 - \eta^{-1} + \varepsilon^2)N_b}{\Delta N(\eta - 1)(1 + \varepsilon^2)}.$$

Let us consider some limiting cases. When $\varepsilon = 0$ and $N_b \neq 0$ we have

$$\frac{\overline{\Delta n^2}}{\bar{n}} = \frac{\eta N_a}{(\eta - 1)\Delta N} - \frac{1}{\eta}.$$

When $N_b = 0$ and with finite detuning we have

$$\frac{\overline{\Delta n^2}}{\bar{n}} = \frac{\eta}{\eta - 1} - \frac{1}{\eta(1 + \varepsilon^2)}.$$

The second term in this expression, connected with the difference between $\tilde{\sigma}$ and σ, becomes small at a large detuning.

In the general case the linewidth $\Delta\nu$ has the form

$$\Delta\nu = \frac{\nu\eta}{4\bar{n}(1 + \varepsilon^2)}\left\{\frac{1 + \varepsilon^2}{\eta} + \frac{N_a + N_b}{\Delta N}\left[(1 + \varepsilon^2)^2 - \frac{2\varepsilon^2(1 + \varepsilon^2)}{\eta} + \frac{\varepsilon^2(\varepsilon^2 - 1)}{\eta^2} + \frac{2\varepsilon^2}{\eta^3}\right]\right\}.$$

APPENDIX II

Let us consider the question of stability of the steady-state oscillatory mode.

The Maxwell self-consistent systems of equations for a classical field $z(t)$ and the equations of motion of a medium with one relaxation time τ have the form

$$\dot{z} + \nu z = -iv; \qquad v = gN\langle\sigma r\rangle$$

$$\dot{v} + v/\tau = igzu; \qquad u = gN\langle\sigma_3 r\rangle$$

55. Brzhazovsky, Yu. V., Vasilenko, L. S., Rautian, S. G., Popova, T. S. and Chebotayev, V. P., *Zh. Experim. i Teor. Fiz.* **61**, 500 (1970).
56. Dicke, R. H., *Phys. Rev.* **93**, 99 (1954).
57. McCumber, D. E., *Phys. Rev.* **141**, 306 (1966).
58. Lax, M., *IEEE J. Quantum Electronics*, **QE-3**, 37 (1967).
59. Hang, H., *Phys. Rev.* **184**, 338 (1969).
60. Surdutovich, G. I. and Titov, E. A., in *Proceedings of the 2nd Vavilov Conference on Nonlinear Optics*, p. 408, Novosibrisk, 1971.
61. Bonifacio, R. and Gronchi, M., *Nuovo Cimento*, **8B**, No. 1, 105 (1972).
62. Arecchi, F. T. and Courtens, E., *Phys. Rev.* **A2**, 1730 (1970).
63. Bonifacio, R., Schwendimann, P. and Haake, F., *Phys. Rev.* **A4**, 403, 854 (1971).
64. Paoli, T. L. and Ripper, J. E., *Phys. Rev.* A2, No. 6, 2551 (1970).
65. Lisitsyn, V. N. and Chebotaev, V. P., *Zh. Eksperim. i Teor. Fiz.* **54**, 419 (1968) (*Sov. Phys.—JETP*, **27**, 227 (1968)).
66. Lee, P. H. and Skolnick, M. L., *Appl. Phys. Lett.* **10**, 303 (1967).
67. Kazantsev, A. P., Rautian, S. G. and Surdutovich, G. I., *Zh. Eksperim. i Teor. Fiz.* **54**, 1409 (1968) (*Sov. Phys.—JETP*, **27**, 756 (1968)).
68. Feld, M. S., Javan, A. and Lee, P. H., *Appl. Phys. Lett.* **13**, 424 (1968).
69. Kazantsev, A. P. and Surdutovich, G. I., *Zh. Eksperim. i Teor. Fiz.* **58**, 245 (1970) (*Sov. Phys.—JETP*, **31**, No. 1, 133 (1970)).
70. Makhnev, V. P. and Telegin, G. G., *Zh. Eksperim. i Teor. Fiz.* **63**, 1212 (1972).
71. Kazantsev, A. P., To be published.
72. Kazantsev, A. P., *Zh. Eksperim. i Teor. Fiz.* **60**, 269 (1971) (*Sov. Phys.—JETP*, **33**, 269 (1971)).
73. Zhelnov, B. L. and Smirnov, V. S., *Zh. Eksperim. i Teor. Fiz.* **57**, 2043 (1969) (*Sov. Phys.—JETP*, **30**, 1108 (1970)).
74. Smirnov, V. S. and Zhelnov, B. L., *Opt. i Spect.* **XXXIII**, No. 3, 505 (1972).
75. Smirnov, V. S. and Tumaikin, A. M., *Opt. i Spect.* **XXXIII**, No. 2, 193 (1972).
76. Zhelnov, B. L. and Smirnov, G. I., *Zh. Eksperim. i Teor. Fiz.* **61**, 1801 (1971).
77. Zaitsev, Yu. I., *Izvestia Vuz'ov Radiofizika*, **XIII**, No. 6, 898 (1970).
78. Grossman, S. and Richter, P. H., *Z. Physik*, **249**, 43 (1971); **255**, 59 (1971).
79. Golubev, Yu. M. and Fradkin, E. E., *Zh. Eksperim. i Teor. Fiz.* **63**, 2082 (1972).
80. Egorov, Yu. P., *Zh. Eksperim. i Teor. Fiz. Pis Red.* **8**, 525 (1968).
81. Stenholm, S., *Physics Reports* (section C of *Physics Letters*), **6**, No. 1, 1 (1973).
82. Spencer, M. B. and Lamb, W. E. Jr., *Phys. Rev.* **A5**, 884 (1972).
83. Kazantzev, A. P., *Zh. Eksperim. i Teor. Fiz.* (in press).
84. Salomaa, R. and Stenholm, S., *Phys. Rev.* **A8** (in press).

$$\dot{u} + (u - gN)/\tau = 2ig(\overset{*}{z}v - \text{c.c.})$$

Linearizing this system of equations near the steady-state solution $u = gN/\eta; |z|^2 = (\eta - 1)/4(g\tau)^2$ and separating the time dependence in the exponential form $e^{tx/\tau}$, we get the following dispersion equation

$$x^3 + ax^2 + bx + c = 0$$

$$a = 2 + \nu\tau; \qquad b = \eta + \nu\tau; \qquad c = 2\nu\tau(\eta - 1).$$

The boundary of the region of instability is obtained from the condition Re $x = 0$, whence it follows that

$$c = ab$$

which agrees with formula (3.16).

APPENDIX III

To take account of the shot noise contribution, let us divide the radiating current j (formula 4.20) into the average and fluctuating parts

$$j = \langle j \rangle + \Delta j, \qquad \langle \Delta j \rangle = 0.$$

The radial and azimuthal diffusion coefficients B' and C' are expressed through the current correlation functions [72]

$$B' = \xi/\nu \,\mathrm{Re} \int_{-\infty}^{t} \mathrm{d}t' \left[\langle \Delta j(t)\, \Delta \overset{*}{j}(t') \rangle - \frac{\overset{*}{z}}{z} \langle \Delta j(t)\, \Delta j(t') \rangle \right]$$

$$C' = \xi/4\nu \,\mathrm{Re} \int_{-\infty}^{t} \mathrm{d}t' \left[\langle \Delta j(t)\, \Delta \overset{*}{j}(t') + \frac{\overset{*}{z}}{z} \langle \Delta j(t)\, \Delta j(t') \rangle \right].$$

In the simplest model of a laser with motionless atoms having one relaxation time τ, these coefficients have the form

$$B' = \xi(\eta - 1)/4\eta; \qquad C' = \xi(\eta - 1)\varepsilon^2/16\eta$$

and the contributions to the dispersion of the photons and the line-width corresponding to them are $\overline{\Delta n^2} = \bar{n}/4$ and $\Delta\nu' = \nu\varepsilon^2(\eta - 1)/8\eta\bar{n}$.

Since the shot noise and quantum fluctuations are statistically independent of each other, the final expressions for the dispersion of photons $\overline{\delta n^2}$ and the linewidth $\delta\nu$ are equal to $\overline{\delta n^2} = \overline{\Delta n^2} + \overline{\Delta n^{2\prime}}$ and $\delta\nu = \Delta\nu + \Delta\nu'$. The contribution to $\overline{\delta n^2}$ from the shot noise becomes appreciable at $\eta - 1 \gtrsim 1$. At $N_b = 0$ and $\eta - 1 \gg 1$ we have $\overline{\delta n^2} = \frac{5}{4}\bar{n}$, i.e. the distribution of the photons differs somewhat from Poisson's law owing to the shot noise.

The contribution of shot noise to phase fluctuations may also be appreciable near the threshold when $\eta - 1 \ll 1$, when the condition $\varepsilon^2(\eta - 1) \gtrsim 1$ holds. When $\varepsilon^2(\eta - 1) \gg 1$ we have

$$\delta\nu \approx \Delta\nu' = \nu\varepsilon^2(\eta - 1)/8\bar{n} = \tfrac{1}{2}(g\tau)^2.$$

Thus, at large detuning near the oscillation threshold, the linewidth is determined only by the saturation parameter.

REFERENCES

1. Haken, H., *Handbuch der Physik*, Vol. XXV/2c. Springer Verlag, Berlin, 1970.
2. Lamb, W. E. Jr., in *Quantum Optics and Electronics* (ed. by C. de Witt, A. Blondin and C. Cohen-Tannoudji), Gordon and Breach, New York, 1965.
3. Lax, M., in *Statistical Physics*, Vol. II (ed. by M. Chretien *et al.*), Gordon and Breach, New York, 1969.
4. Klimontovich, Yu. L., *Uspekhi Fiz. Nauk*, **101**, 577 (1970).
5. Fleck, J. A. Jr., *Phys. Rev.* **149**, 309, 322 (1966).
6. Willis, C. R., *Phys. Rev.* **147**, 406 (1966); **156**, 320 (1967).
7. Scully, M. O. and Lamb, W. E. Jr., *Phys. Rev.* **159**, 208 (1967).
8. Lax, M., *Phys. Rev.* **172**, No. 2, 362 (1968) and references quoted therein.
9. Gordon, J. P., *Phys. Rev.* **161**, 367 (1967).
10. Haken, H. and Weidlich, W., Lectures at the Intern. School of Physics "Enrico Fermi", XLII Course, Quantum Optics, Academic Press, New York and London, 1969.
11. Kazantsev, A. P. and Surdutovich, G. I., *Zh. Eksp. Teor. Fiz.* **56**, 2001 (1969) (*Sov. Phys.—JETP*, **31**, 133 (1970)).
12. Notkin, G. E., Rautian, S. G. and Feoktistov, A. A., *Sov. Phys.—JETP*, **25**, 1112 (1967).
13. Rautian, S. G., *Tr. Fiz. Inst. Akad. Nauk*, **35**, 43 (1968).
14. Stroud, C. R. Jr., *Phys. Rev.* **A3**, 1044 (1971).
15. Schwinger, J., *J. Math. Phys.* **2**, No. 3, 407 (1961).
16. Glauber, R. T., *Phys. Rev.* **131**, 2766 (1963).
17. Klauder, J. R. and Sudarshan, E. C. G., *Fundamentals of Quantum Optics*, New York, Amsterdam, 1968.
18. Louisell, W. H., *Radiation and Noise in Quantum Electronics*, New York, 1964.
19. Mandel, L. and Wolf, E., *Phys. Rev.* **149**, 1033 (1966).
20. Scully, M. O. and Lamb, W. E. Jr., *Phys. Rev.* **166**, 246 (1968).

21. Fain, V. M. and Khanin, Ya. I., *Quantum Electronics*, Vol. I, Pergamon Press, Oxford, 1967.
22. Kazantsev, A. P. and Smirnov, V. S., *Zh. Eksperim. i Teor. Fiz.* **46**, 182 (1964) (*Sov. Phys.—JETP* **19**, 130 (1964)).
23. Alekseev, A. I., Vdovin, Yu. A. and Galitskii, V. M., *Zh. Eksperim. i Teor. Fiz.* **46**, 320 (1964) (*Sov. Phys.—JETP*, **19**, 220 (1964)).
24. Arecchi, F. T., Lectures at the Intern. School of Physics "Enrico Fermi", XLII Course, Quantum Optics, Academic Press, New York and London, 1969.
25. Hauz, H. A., Lectures at the Intern. School of Physics "Enrico Fermi", XLII Course, Quantum Optics, Academic Press, New York and London, 1969.
26. Armstrong, J. A. and Smith, A. W., *Progress in Optics*, **6**, 213 (1967).
27. Chang, R. F., Korenman, V., Alley, C. O. and Detenbeck, R. W., *Phys. Rev.* **178**, No. 2, 612 (1969).
28. Glauber, R. J., in *Quantum Optics and Electronics* (ed. by C. de Witt, A. Blondin and C. Cohen-Tannoudji), Gordon and Breach, New York, 1965.
29. Glauber, R. J., Lectures at the Intern. School of Physics "Enrico Fermi", XLII Course, Quantum Optics, Academic Press, New York and London, 1969.
30. Scully, M. O. and Lamb, W. E., Jr., *Phys. Rev.* **179**, No. 2, 368 (1969).
31. Arecchi, F. T. and Degiorgio, V., *Phys. Rev.* **3A**, 1108 (1971).
32. Steer, W. H., *Proc. IEEE*, **54**, 11, 1966.
33. Baklanov, E. V., Rautian, S. G., Troshin, B. I. and Chebotaev, V. P., *Zh. Eksperim. i Teor. Fiz.* **56**, 1120 (1969) (*Sov. Phys.—JETP*, **29**, 601 (1969)).
34. Telegin, G. G., Dissertation, Novosibirsk, 1973.
35. Risken, H. and Vollmer, H. D., *Z. Physik*, **204**, 240 (1967).
36. Gordon, J. P. and Aslaksen, T. W., *IEEE J. Quantum Electronics*, **QE-6**, 7 (1970).
37. Sargent, M., Scully, M. and Lamb, W. E. Jr., *Appl. Optics*, **9**, 11 (1970).
38. Zaitsev, Yu. I. and Stepanov, D. P., *ZhETF Pis. Red.* **6**, 733 (1967) (*JETP Lett.* **6**, 209 (1967)).
39. Zaitsev, Yu. I. and Stepanov, D. P., *Zh. Eksperim. i Teor. Fiz.* **55**, 1645 (1968).
40. Arrathoon, R. and Siegman, A. E., *J. Appl. Phys.* **40**, No. 2, 910 (1969).
41. Arrathoon, R. and Siegman, A. E., *Phys. Rev. Lett.* **20**, 901 (1968).
42. Risken, H. and Vollmer, H. D., *Z. Physik*, **201**, No. 3, 323 (1967).
43. Risken, H., *Fort. der Phys.* **16**, 5, 261 (1968).
44. Davidson, F. and Mandel, L., *Phys. Lett.* **27A**, 579 (1968).
45. Cantrell, C. D., Lax, M. and Smith, W. A., *Phys. Rev.* **7A**, No. 1, 175 (1973).
46. Chopa, S. and Mandel, L., *Phys. Rev. Lett.* **30**, No. 2, 60 (1973).
47. Andronova, I. A. and Zaitsev, Yu. I., *Izvestia Vuz'ov, Radiofizika*, **11**, No. 1, 149 (1968).
48. Andronova, I. A., *Zh. Eksperim. i Teor. Fiz.* **56**, No. 2, 417 (1969).
49. Zaitsev, Yu. I., *Izvestia Vuz'ov, Radiofizika*, **13**, 898 (1970).
50. Haldre, H. Yu. and Kokhlov, R. V., *Izvestia Vuz'ov Radiofizika*, **1**, No. 5–6, 60 (1958).
51. Gurtovnik, A. S., *Izvestia Vuz'ov, Radiofizika*, **1**, No. 5–6, 83 (1958).
52. Korobkin, V. V. and Uspenskii, A. V., *Zh. Eksperim. i Teor. Fiz.* **45**, 1003 (1963) (*Sov. Phys.—JETP*, **18**, 693 (1963)).
53. Zhelnov, B. L., Kazantsev, A. P., Kolpashchikov, V. L. and Smirnov, V. S., *Fiz. Tverd. Tela*, **7**, 1756 (1965) (*Sov. Phys.—Solid State*, **7**, 1416 (1965)).
54. Kazantsev, A. P., *Zh. Eksperim. i Teor. Fiz.* **61**, 1790 (1971) (*Sov. Phys.—JETP*, **34**, 953 (1972)).